CliffsQuickR[illegible]

Astronomy

By Charles J. Peterson, Ph.D.

CLIFFS QUICK REVIEW™ Astronomy
Published by
Course Hero

www.coursehero.com
www.cliffsnotes.com (CliffsNotes Website)

Library of Congress Control Number: 00-103369
ISBN: 978-1-957671-31-4

CONTENTS

CONTENTS

CONTENTS

CONTENTS

CHAPTER 1
A BRIEF HISTORY OF ASTRONOMY

Astronomy, which literally means "the study of the stars," involves the study of the universe and every type of object that exists in the universe, including our own planet. **Astronomers** study the universe by investigating the origins, evolution, composition, motions, relative positions, and sizes of celestial bodies.

Unlike **astrology,** which alleges that the positions of the Sun, Moon, and stars affect human affairs, astronomy is a scientific discipline based upon an understanding of physical principles (the so-called Laws of Nature) and how their operation produces phenomena that may be observed. These principles can often be summarized in simple mathematical equations. Learning some of the simpler and more fundamental principles allows you both insight into how astronomers learn about the universe as well as an understanding of why the objects we observe have the properties that they do.

The number of physical factors that astronomers can measure is small. These factors include masses, sizes, densities, temperatures, and brightnesses, as well as the time over which changes may occur. Often, these factors are measured in the metric system based on kilograms, meters, and seconds (see Chapter 2). Many quantities can be numerically large; therefore, astronomers also use other units specific to the discipline of astronomy, such as astronomical units for distances within the solar system, parsecs for distances to stars, solar masses (the mass of the Sun) for masses of stars, and solar luminosities (the luminosities of the Sun) for the energy-per-second emitted by stars (see Chapter 8).

Astronomy has had a long, rich history, from ancient peoples' interpretations of celestial phenomena to modern studies of the universe. Many individuals have made significant contributions to the development of astronomy throughout human history.

Archaeoastronomy

Archaeoastronomy is the study of ancient (pre-technological) humankind's awareness of celestial phenomena and its influence on their societies. A number of these influences in ancient societies have been identified, including how structures were built, early forms of calendars, and the development of mathematical concepts.

Ancient structures

Ample evidence exists that in ancient times, many cultural groups all over the world built single structures, buildings, and even cities facing in astronomically significant directions. For example, the Newgrange passage grave in Ireland (circa 3200 B.C.) faces the mid-winter sunrise, and Stonehenge, in southern England (circa 2800 B.C.), points toward the mid-summer sunrise. Similarly, various Mayan buildings in Mesoamerica show relationships to significant horizon rise and set positions for the planet Venus.

Early calendars

Before ancient humans recognized that the yearly cycle predictably repeats itself, they regulated agricultural, hunting, and religious activities by directly observing celestial phenomena. These phenomena included solar position against the background of stars, the cyclic north-south oscillation of the Sun's rise and set positions on the horizon, or the pattern of nighttime constellations. Evidence of early peoples' celestial timekeeping include the orientation of Stonehenge, which suggests a calendrical awareness, and the Sun Circle (circa 1100 A.D.) built by the people of the Indian city of Cahokia, Illinois.

Navigation

When societies embarked on extended travel where natural features on land were not available, the sky provided an alternative means to mark one's position during journeys. The Polynesians were experts at using the apparent positions of stars in the sky to navigate across the vast expanses of the southern Pacific Ocean. In the same manner, traders crossing the featureless tracks of the Sahara Desert used the sky for their directions. It has been argued that our oldest constellations may date from the delineation of stellar patterns as mnemonic devices for Mediterranean navigation as early as 3000 B.C.

Mathematics

In due course, as societies accumulated extensive observations of the sky, there developed the desire to understand celestial events in a quantitative fashion. The 360 degrees of the circle is the most tractable approximation that the Babylonians could make for the annual circuit of the Sun around the sky in 365.25 days. Islamic scholars were inspired to develop spherical trigonometry for navigational reasons as well as for the Islamic tenet to pray toward the city of Mecca.

Other cultural influences

Awareness of phenomena in the sky has affected other aspects of cultures. Descendancy from the celestial deities of the Sun and Moon was used in many past societies to justify political supremacy. The development of mythologies, astrologies, and religions also contain elements that help societies obtain a sense of order regarding the workings of celestial events.

Greek Astronomy

The greatest influence on western astronomy comes from ancient Greece. Greek philosophers were the first to apply mathematics to attempt to understand the universe more deeply than for simple predictive purposes. Although the civilization of ancient Greece declined, the Greek ideas that were passed to Indian and Asian colonies were subsequently adopted by Islamic scholars, and were later introduced into Europe by Arab invaders in the fifteenth century Table 1-1 gives a listing of the Greek alphabet. These letters note scientific quantities and are used in the naming of stars.

Table 1-1: The Greek Alphabet

Letter	*Upper Case*	*Lower Case*	*Letter*	*Upper Case*	*Lower Case*
Alpha	Α	α	Nu	Ν	ν
Beta	Β	β	Xi	Ξ	ξ
Gamma	Γ	γ	Omicron	Ο	ο
Delta	Δ	δ	Pi	π	π
Epsilon	Ε	ε	Rho	Ρ	ρ
Zeta	Ζ	ζ	Sigma	Σ	σ
Eta	Η	η	Tau	Τ	τ
Theta	Θ	θ	Upsilon	Υ	υ
Iota	Ι	ι	Phi	Φ	φ
Kappa	Κ	κ	Chi	Χ	χ
Lambda	Λ	λ	Psi	ψ	Ψ
Mu	Μ	μ	Omega	Ω	ω

Pythagoras

Pythagoras (circa 580–500 B.C.), who is credited with summarizing previously developed ideas of geometry, is also credited with proposing the idea of a spherical Earth and Moon. He developed this concept from studying the pattern of shadows on the Moon during eclipses. Pythagoras also correctly deduced the cause of the phases of the Moon.

Aristotle

Aristotle (circa 384–322 B.C.) believed that the Sun moved around Earth. His belief in a stationary Earth was probably the result of his familiarity with the concept of parallax and his interpretation of observed celestial movement. **Parallax** refers to the apparent change in the position of an object resulting from the change in the direction or position from which it is viewed. Understanding parallax, Aristotle would have expected that if Earth rotated and moved about the Sun, then parallax should be observable for nearby celestial objects (see Chapter 8). His inability to observe the parallax effect led to the conclusion that Earth is stationary, rather than the correct deduction that the objects are so far away that the parallax angles are too small to detect. Aristotle also argued for the philosophical concepts of perfect celestial shapes (spheres) and motions (uniform on circular paths).

Aristarchus

Aristarchus (fl. circa 270 B.C.) used a trigonometric interpretation of a few simple observations to develop the first valid idea of the scale of the universe. He first noticed that the apparent angular sizes of the Sun and Moon are the same, hence their sizes (radii) are in proportion to their distances. Second, he estimated that the angle between the direction to the quarter Moon and the Sun is 87 degrees. Recognizing that the Sun-Moon-Earth angle must be a right angle (90 degrees) when the Moon is exactly half illuminated, Aristarchus determined that the distance from Earth to the Sun is 20 times that to the Moon. Third, by timing a lunar eclipse, Aristarchus could establish that three moon diameters fit within the diameter of Earth's shadow. Fourth, the ratio of the eclipse time to the lunar orbital period gives the lunar size relative to its orbital distance [eclipse time/orbital period = 3 diameter (moon)/2 π (moon)]. Taken together, these observations place the Moon at a distance of 10 Earth diameter (as compared to the true value of 30 Earth diameter) and the Sun at 200 Earth diameter (true value 11,700), with the Moon one-third (true value 0.27 times the size of Earth) and the Sun seven times the size of Earth (true value 109 times the size of the Earth). Although his values differ from accepted values, his effort was a significant advance. Aristarchus was also the only significant Greek philosopher to believe that Earth moved, although he ultimately discarded this idea.

Eratosthenes

Eratosthenes (circa 276–194 B.C.) determined the circumference of Earth (40,000 kilometers) from observations at Alexandria and Syene, Egypt, by noting that the seven-degree angle between the direction to the Sun at these two sites on the same date is the same fraction of a full circle (360 degrees) as is the distance between the two towns to the circumference of Earth.

Hipparchus

The work of Hipparchus (fl. 146–127 B.C.), often thought of as the greatest of the ancient Greek astronomers, shows the transition from basic naked-eye astronomy to systematic observational work with a long series of written records being kept. From long-term records, he could determine the length of a year to an accuracy of 6 minutes and note that the seasons were of unequal length, now known to be an effect of the ellipticity of Earth's orbit. He was successful in the prediction of eclipses. Comparison of his observations of the sky with the writings of Eudoxus three centuries earlier led to the realization that Earth's rotation axis had moved, an effect known as **precession.** (As it spins, Earth is like a top for which the rotational axis moves about in a circle.)

Hipparchus was also the first to compile a catalog of the 850 brightest stars, with a numerical estimate of their brightness estimates. In Hipparchus's scale, the brightest naked-eye stars were designated category 1. Somewhat fainter stars were designated 2, still fainter 3, and so forth, to 6, the faintest naked-eye stars. These categories, called **magnitudes** in modern terminology, are based on the physiological response of the eye and brain, which perceives brightness *ratios*. On average, a magnitude 1 star is about 2.5 times as bright as a magnitude 2 star, a magnitude 2 star is 2.5 times as bright as a magnitude 3 star, and so forth. Magnitudes are still used as a measurement of brightness. The scale is extended to brighter objects (Sun, full Moon) by going to negative numbers (–26, –12.5, respectively) and to fainter objects by going to larger numerical magnitudes (the largest telescope with the best light detection devices and integration over time can detect stars as faint as magnitude 30).

Ptolemy and the Geocentric hypothesis

The writings of Ptolemy (fl. 140 A.D.), the last of the ancient Greek astronomers, show continued numerical advance over his predecessors. By his era, for example, trigonometric and geometric analysis of lunar observations had refined the Moon's distance to essentially the value accepted today.

Ptolemy attempted to explain what astronomers of his era could see occurring in the sky (see Chapter 3) in terms of a **geocentric model,** at the center of which was a stationary Earth. He first assumed that the whole system revolved once per day around a stationary Earth. Each of the planets moved at uniform rates on small circles **(epicycles)**, which in turn moved uniformly around larger circles (**deferents**), with the center of each deferent offset slightly from the position of Earth. The Sun and Moon, however, used no epicycles (both have almost uniform motion about the sky). For Mercury and Venus, Ptolemy needed to lock the centers of their epicycles to the direction of the Sun. The position of each outer planet on its epicycle was such that the center-planet direction is parallel to the Earth-Sun direction. Finally, the distances of all celestial objects must be based on the apparent motions relative to the stars and on sizes of the retrograde loops, thus the order outward from Earth is the Moon, Mercury, Venus, the Sun, Mars, Jupiter and Saturn. The stars were placed exterior to Saturn.

Foundations of Modern Astronomy

Nicolaus Copernicus and the Heliocentric hypothesis

Copernicus (1473-1547) was a Polish scholar who postulated an alternative description of the solar system. Like the Ptolemaic geocentric ("Earth-centered") model of the solar system, the Copernican **heliocentric** ("Sun-centered") **model** is an **empirical model.** That is, it has no theoretical basis, but simply reproduces the observed motions of objects in the sky.

In the heliocentric model, Copernicus assumed Earth rotated once a day to account for the daily rise and set of the Sun and stars. Otherwise the Sun was in the center with Earth and the five naked-eye planets moving about it with uniform motion on circular orbits (deferents, like the geocentric model of Ptolemy), with the center of each offset slightly from Earth's position. The one exception to this model was that the Moon moved about Earth. Finally, in this model, the stars lay outside the planets so far away that no parallax could be observed.

Why did the Copernican model gain acceptance over the Ptolemaic model? The answer is not accuracy, because the Copernican model is actually no more accurate than the Ptolemaic model—both have errors of a few minutes of arc. The Copernican model is more attractive because the principles of geometry set the distance of the planets from the Sun. The greatest angular displacements for Mercury and Venus (the two planets that orbit closer to the Sun, the so-called **inferior** planets) from the position of the Sun (**maximum elongation**) yield right angle triangles that set their orbital sizes relative to Earth's orbital size. After the orbital period of an outer planet (a planet with an orbital size larger than the orbit of Earth is termed a **superior** planet) is known, the observed time for a planet to move from a position directly opposite the sun (**opposition**) to a position 90 degrees from the Sun (**quadrature**) also yields a right-angle triangle, from which the orbital distance from the Sun can be found for the planet.

If the Sun is placed in the center, astronomers find that planetary orbital periods correlate with the distance from the Sun (as was *assumed* in the geocentric model of Ptolemy). But its greater simplicity does not prove the correctness of the heliocentric idea. And the fact that Earth is unique for having another object (the Moon) orbiting around it is a discordant feature.

Galileo Galilei and the invention of the telescope

Settling the debate between the geocentric versus heliocentric ideas required new information about the planets. Galileo did not invent the telescope but was one of the first people to point the new invention at the sky, and is certainly the one who made it famous. He discovered craters and mountains on the Moon, which challenged the old Aristotelian concept that celestial bodies are perfect spheres. On the Sun he saw dark spots that moved about it, proving that the Sun rotates. He observed that around Jupiter traveled four moons (the **Galilean satellites** Io, Europa, Callisto, and Ganymede), showing that Earth was not unique in having a satellite. His observation also revealed that the Milky Way is composed of myriads of stars. Most crucial, however, was Galileo's discovery of the changing pattern of the phases of Venus, which provided a clear-cut test between predictions of the geocentric and heliocentric hypotheses, specifically showing that the planets must move about the Sun.

Johannes Kepler

Because the heliocentric concept of Copernicus was flawed, new data were required to correct its deficiencies. Tycho Brahe's (1546-1601) measurements of accurate positions of celestial objects provided for the first time a continuous and homogeneous record that could be used to mathematically determine the true nature of orbits. Johannes Kepler (1571-1630), who began his work as Tycho's assistant, performed the analysis of planetary orbits. His analysis resulted in **Kepler's laws of planetary motion,** which are as follows (see also Chapter 2):

The law of orbits: All planets move in elliptical orbits with the Sun at one focus.

The law of areas: A line joining a planet and the Sun sweeps out equal areas in equal time.

The law of periods: The square of the period (P) of any planet is proportional to the cube of the semi-major axis (r) of its orbit, or P^2G (M (sun) + M) = $4\pi^2 r^3$, where M is the mass of the planet.

Isaac Newton. Isaac Newton (1642-1727), in his 1687 work, *Principia,* placed physical understanding on a deeper level by deducing a law of gravity and three general laws of motion that apply to all objects (see also Chapter 2):

Newton's first law of motion states that an object remains at rest or continues in a state of uniform motion if no external force acts upon the object.

Newton's second law of motion states that if a net force acts on an object, it will cause an acceleration of that object.

Newton's third law of motion states that for every force there is an equal and opposite force. Therefore, if one object exerts a force on a second object, the second exerts an equal and oppositely directed force on the first one.

Beyond Newton: Relativity Theories

Newton's Laws of Motion and Gravity are adequate for understanding many phenomena in the universe; but under exceptional circumstances, scientists must use more accurate and complex theories. These circumstances include **relativistic conditions** in which a) large velocities approaching the speed of light are involved (theory of **special relativity**), and/or b) where gravitational forces become extremely strong (theory of **general relativity**).

In simplest terms, according to the theory of general relativity, the presence of a mass (such as the Sun) causes a change in the geometry in the space around it. A two-dimensional analogy would be a curved saucer. If a marble (representing a planet) is placed in the saucer, it moves about the curved rim in a path due to the saucer's curvature. Such a path, however, is the same as an orbit and nearly identical with the path that would be calculated by use of a Newtonian gravitational force to continually change the direction of motion. In the real universe, the difference between Newtonian and relativistic orbits is usually small, a difference of two centimeters for the Earth-Moon orbital distance (r = 384,000 km on average).

CHAPTER 2
THE SCIENCE OF ASTRONOMY

The science of the universe, like any other natural science, is an organized study of natural phenomena. Consequently, in order to study the concepts of astronomy, one must first be familiar with some basic scientific principles.

The Scientific Method

An area of inquiry is a scientific discipline if its investigators use the **scientific method**—a systematic approach to researching questions and problems through objective and accurate observation, collection and analysis of data, direct experimentation, and replication of these procedures. Scientists emphasize the importance of gathering information carefully and accurately, and researchers strive to remain unbiased when evaluating information, observing phenomena, conducting experiments, and recording procedures and results. Researchers also recognize the value of skepticism and the necessity of having their findings confirmed by other scientists.

The scientific method is an idealization of the process by which scientific understanding advances. A scientist starts with what is known about a natural phenomenon (for example, the data) and will develop a **theory,** or an integrated set of statements, that explain various phenomena. Because a theory is too general to test, the investigator devises a **hypothesis**—a testable prediction—from the theory and tests the hypothesis instead of a general theory. If the hypothesis passes the test, then its acceptance as an accurate description of physical phenomena is strengthened. Repeated testing may raise the hypothesis to the status of a **Law of Nature**, a well-confirmed summary statement of how a natural phenomenon behaves. If the hypothesis is disproved, then it may need to be revised or even to be discarded. Chance discovery of new information may also require revision of a hypothesis.

Measurement Methods

Astronomers use a variety of methods to quantify the natural phenomena that they measure. Because the physical factors that are measured in astronomy are sometimes quite large or small, astronomers use **scientific notation,** or exponential numbers, to express these measurements. A number written in scientific notation is a number between 1 and 10 and multiplied by a power of 10 (see Table 2-1).

Table 2-1: Scientific Notation and Powers of Ten

Number Name	*Number*	*Scientific (Exponential) Form*
trillion	1,000,000,000,000	10^{12}
billion	1,000,000,000	10^{9}
million	1,000,000	10^{6}
thousand	1,000	10^{3}
hundred	100	10^{2}
ten	10	10^{1}
one	1	10^{0}
one-tenth	$^{1}/_{10}$	10^{-1}
one-hundreth	$^{1}/_{100}$	10^{-2}
one-thousandth	$^{1}/_{1000}$	10^{-3}
one-millionth	$^{1}/_{1,000,000}$	10^{-6}
one-billionth	$^{1}/_{1,000,000,000}$	10^{-9}

Astronomers measure factors such as size or distance using the **metric system,** a decimal system of weights and measures in which the kilogram (2.04 pounds), the meter (39.37 inches), and the liter (61.025 cubic inches) are the basic units of mass, length, and capacity, respectively. Names for the most common other units are formed by the addition of the following prefixes to these terms: kilo- (thousand), mega- (million), giga- (billion), centi- ($^1/_{100}$), milli- ($^1/_{1000}$), and micro ($^1/_{1,000,000}$). So, for example, a kilometer is a thousand meters, while a millimeter is one-thousandth of a meter. Some astronomical measurements using the metric system are shown in Table 2-2.

Table 2-2: Distances

Units	***Examples***
1 centimeter (cm) = 10 millimeters (mm) = 0.394 inches	Thickness of pencil: 7 mm Diameter of golf ball: 4 cm
1 meter (m) = 100 cm = 1,000 mm = 39.4 inches = 1.09 yards	Basketball player: 2 m Football field: 90 m
1 kilometer (km) = 1,000 m = 1.609 miles	United States: 4,900 km across Distance to Moon: 384,000 km
1 Astronomical Unit (AU) = 1.50×10^8 km = 9.30×10^7 miles	Earth to Sun: 1 AU Sun to Pluto: 39 AU
1 light-year (ly) = 63,270 AU = 9.5×10^{12} km	Nearest star: 4 ly Diameter of Galaxy: 80,000 ly
1 parsec (pc) = 3.26 ly = 206,265 AU = 3.09×10^{13} km = 3.09×10^{18} cm	Nearest star: 1.3 pc Nearest large galaxy: 690,000 pc

Astronomers also utilize angles in measuring celestial objects, especially in quantifying their position and movement in the sky. An **angle** is formed by two lines that have the same endpoint. An angle is measured in degrees, from 0 to 180. The number of degrees indicates the size of the angle. A **right angle** has a measure of 90 degrees, while a complete circle would measure 360 degrees (see Figure 2-1). Angular measure is often given in terms of arc, which can be measured in minutes (60 minutes to a degree) or seconds (60 seconds to a minute).

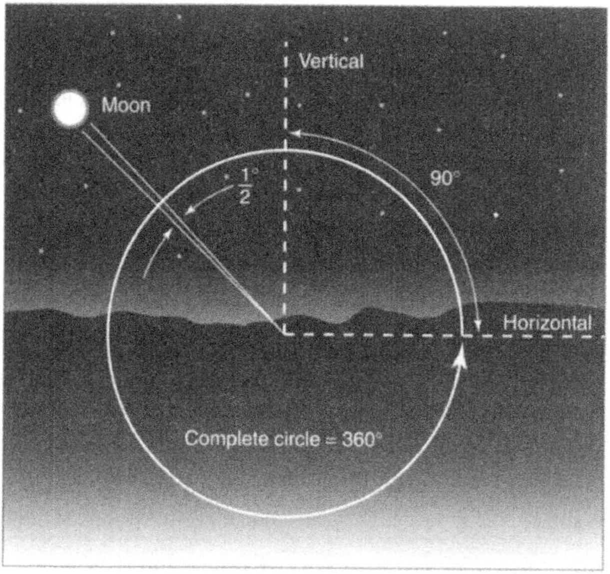

Figure 2-1

Measuring angles. Notice that the Moon's angular diameter can be measured as ½ degree or 30 minutes of arc.

Basic Physics

In the sixteenth and seventeenth centuries, scientists discovered the laws of the motion of material objects. These laws help scientists to explain and predict the motions of celestial bodies.

Kepler's Three Laws of Planetary Motion

As mentioned in Chapter 1, Kepler formulated three laws to approximate the behavior of planets in their orbits. To understand Kepler's First Law of Planetary Motion (**Law of Ellipses**), one must first be familiar with the properties and components of an **ellipse.** An ellipse is the path of a point that moves so that the sum of its distances from two fixed points (the foci) is constant. An ellipse has two axes of symmetry. The longer one is called the **major axis,** and the shorter one is called the **minor axis.** The two axes intersect at the center of the ellipse. Kepler's First Law states that the orbit of a planet is an ellipse with the Sun at one focus (see Figure 2-2). The size of an ellipse is given by the length of the **semi-major axis** (half of the major axis), which is also equal to the average distance of the planet to the Sun as it travels about the Sun in its orbit. The shape of an ellipse is measured by the **eccentricity,** or a measure of how much an ellipse deviates from the shape of a circle ($e = CF/a = (1 - b^2/a^2)^{1/2}$). Therefore, a circle would have an eccentricity of 0, while a line would have an eccentricity of 1. The closest approach of a planet to the Sun is known as the **perihelion**, a distance equal to $a(1 - e)$. The greatest distance between a planet and the Sun is the **aphelion** equal to $a(1 + e)$ (see Figure 2-3).

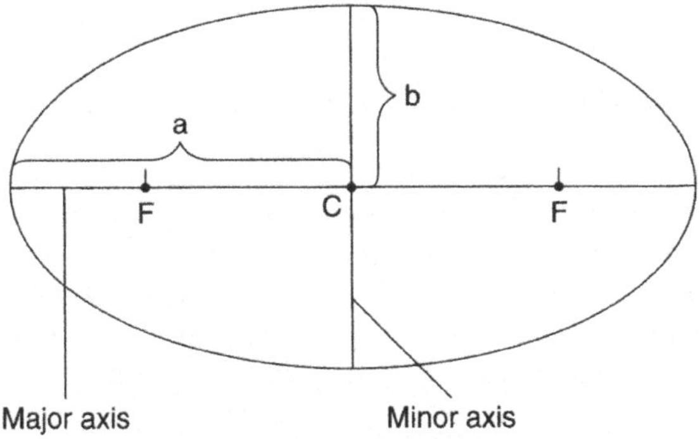

Figure 2-2

An ellipse. The shape of the ellipse is determined by the ratio of the distance between the two foci (F) to the length of the major axis (the eccentricity). If the foci are closer together, the ellipse will have a smaller eccentricity and will more closely resemble a circle. If the foci are farther apart, they will have a greater eccentricity and will more closely resemble a straight line.

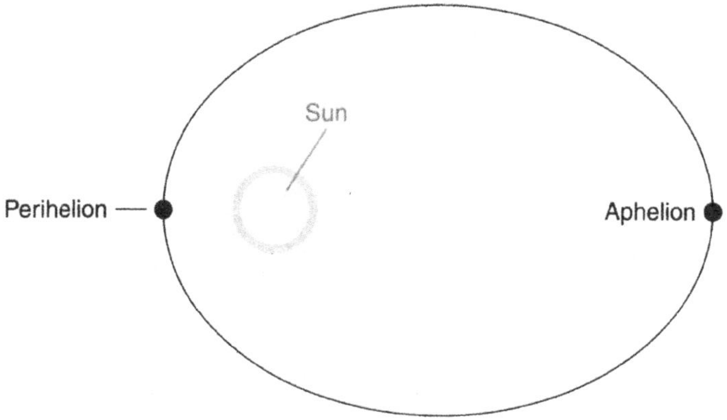

Figure 2-3

The elliptical orbit of a planet around the Sun (ellipticity is greatly exaggerated; most orbits are nearly circular).

Kepler's Second Law of Planetary Motion (**Law of Areas**) states that a line connecting the planet with the Sun sweeps over an area at a constant rate. In other words, if the time for an object to move from position A to position B is the same as the time to move from C to D, the areas swept out are also equal. This law is actually an alternative statement of the physical principle of **the conservation of angular momentum**: In the absence of an outside force, angular momentum = mass × orbital radius × the tangential velocity (that is, the velocity perpendicular to the radius) does not change. In consequence, when a planet moves closer to the Sun, its orbital velocity must increase, and vice versa.

Kepler's Third Law of Planetary Motion (**Harmonic Law**) details an explicit mathematical relationship between a planet's orbital period and the size of its orbit, a correlation noted by Copernicus. Specifically, the square of a planet's period *(P)* of revolution about the Sun is proportional to the cube of its average distance *(a)* from the Sun. For example, P^2 = constant a^3. If P is expressed in years and the semi-major axis a in astronomical units, the constant of proportionality is 1 yr^2/AU^3, and the proportionality becomes the equation $P^2 = a^3$.

Although Kepler's Laws were deduced explicitly from study of planets, their description of orbital properties also applies to satellites moving about planets and to situations in which two stars, or even two galaxies, move about each other. The Third Law, in the form as proposed by Kepler, however, applies only to planets whose masses are negligible in comparison to that of the Sun.

Newton's Three Laws of Motion and a Law of Gravitation

Newton's First Law of Motion (**Law of Inertia**) states that an object continues moving at the same rate unless acted upon by an outside (external) force. If no external force interferes, a moving object keeps moving at a constant **velocity** (that is, both speed and direction remain the same). Similarly, an object at rest stays at rest. This tendency of matter to remain at rest if at rest, or, if moving, to keep moving in the same direction at the same speed is called **inertia**. **Mass** is what gives an object inertia. Mass is a measure of the quantity of material in an object, not its weight, which is a measure of the gravitational force exerted on an object. Newton's First Law is a statement of the modern principle of **Conservation of Momentum,** where **momentum** *(p)* is an object's mass *(m)* times its velocity *(v)*. Momentum stays constant if the outside force is zero.

As in any mathematical expression of a physical law, each term has a precise definition and meaning. Both velocity and momentum are **vector quantities**; that is, each has both a size and a direction. Thus, $p = mv$ involves both the magnitudes of the quantities involved and their directions; momentum and velocity are expressed as boldface (or sometimes with an arrow above the symbol) to remind the user that a direction is involved. This physical law is not something one would intuitively derive from real-world observations, for virtually no real circumstances exist without outside forces—usually friction—acting on objects. Friction can't be seen; therefore, one tends to forget about it; but it is a real force.

Newton's Second Law of Motion (**Law of Force**) states that if a force acts upon an object, the object accelerates in the direction of the force, its momentum changing at a rate equal to that force. **Force F** is the agent that causes a change in a body's momentum. **Force** = rate of change of **momentum** = rate of change of (mass × **velocity**) = mass × rate of change of **velocity** = mass × **acceleration**; or, more familiarly, **F** = m**a** where again the boldface indicates the vector nature of both force and the acceleration. Newton's first law is a direct consequence of the second: If no force acts, there is no acceleration. No acceleration (a change in velocity divided by the time over which the change occurred) means no change in the velocity.

The combination of the operation of these two laws is sufficient to explain orbital motion. If left alone, an object continues its motion in a straight line. Application of a gravitational force on the object produces an acceleration in the direction of the force, which also produces a velocity component in the direction of the force. The combination of the two motions produces a velocity in a new direction, along which the object continues to move as long as no other force is introduced. As a result, an object like the Moon, being acted upon by Earth's gravitational force, literally "falls" around a larger object like Earth.

Newton's Third Law of Motion (**Law of Reaction**) states that forces always occur in mutually acting pairs. In other words, forces are reciprocal; for every force, there is an equal and opposite force.

Newton's Third Law has a significant consequence for Kepler's Third Law of Planetary Motion, which was derived from the assumption that the Sun is stationary. In actuality, the Sun feels a gravitational force due to the pull of the planet, and the planet feels a gravitational force due to the pull of the Sun. From the Second Law, acceleration = force / mass. The smaller mass (the planet) experiences the greater acceleration, hence the greater resultant velocity, and therefore the larger orbit. The Earth/Moon mass ratio is $^{81}/_{1}$, thus the Moon's orbit is 81 times that of Earth's orbit around the **common**

center of mass, the balance point along a line joining two objects, of the Earth/Moon system. The Earth/Sun mass ratio is 1/330,000; thus, Earth's orbit is 330,000 times bigger than the Sun's orbit. Letting the mass of any two objects in orbit about each other be M_1 and M_2, then the common center of mass is determined by $M_1a_1 = M_2a_2$ where $a_1 + a_2 = a$, the relative semi-major axis as used by Kepler. Because of Newton's law of reciprocity, Kepler's Third Law must be rewritten: $P^2(M_1 + M_2) = a^3$ where it is assumed that masses are measured in solar masses, orbital periods in years, and the relative orbital semi-major axis in astronomical units (expressed in other units, the general form of Kepler's Third Law is $P^2G(M_1 + M_2) = 4\pi^2a^3$ where the gravitational constant *(G)* must be given an appropriate value). As modified by Newton, Kepler's Third Law becomes an indispensable tool for determining the masses of other objects in the universe whose orbital motions may be measured.

Newton's Law of Gravitation states that between any two objects there exists a force of attraction proportional to the product of their masses and inversely proportional to the square of the distance between them. Therefore, $F = G\, m_1m_2/r^2$. The Law of Gravity is a universal law that applies to everything including all phenomena on Earth, the motions of the planets, motions of the stars in the Galaxy, motions of the galaxies in the great clusters of galaxies, and everything else in the universe. Gravity is the dominant force in the macroscopic (large-scale) universe, but it is actually the weakest of the four known forces in nature (the other three are the electromagnetic, the strong nuclear, and the weak nuclear forces). Although it is a mathematically simple law, the fact that the strength of gravity depends on distance makes application of the law to most real circumstances exceedingly difficult to apply; even a small change in position results in a change in force, thus a change in the acceleration or rate of change of velocity. This means that velocity and position cannot be written as simple algebraic expressions, but must be expressed as summations of many small, ever different, changes. Newton was forced to invent calculus in order to compute orbits; but with this new mathematical formulism, he was able to show that orbits are indeed described by Kepler's Laws.

Electromagnetic Radiation (Light)

The second great area of physics necessary to address the universe is the subject of **light**, or **electromagnetic radiation**. **Visible light** is the relatively narrow frequency band of electromagnetic waves to which our eyes are sensitive. Wavelengths are usually measured in units of nanometers (1 nm = 10^{-9}m) or in units of **angstroms** (1 Å = 10^{-10}m). The colors of the visible spectrum stretch from violet with the shortest wavelength to red with the longest wavelength.

However, electromagnetic radiation consists of more than just visible light; it also includes (from short wavelength to long wavelength) gamma-radiation, X-radiation, ultraviolet, visible, infrared (heat), microwaves, and radio waves (see Figure 2-4). All of these forms of light have both electrical and magnetic characteristics. The properties of light (see the section, "Particle properties of light") allow us to build devices to observe the universe and to deduce the physical nature of the sources that emit the radiation received during these observations. However, these same properties mean that light interacts with other matter before it reaches the observer and this often complicates our ability to observe other objects in the universe. Note that the word "radiation" can refer to any phenomena that radiates (moves) outwards from a source, here electromagnetic or light radiation. The term should not be confused with radiation associated with a radioactive source, i.e. nuclear radiation.

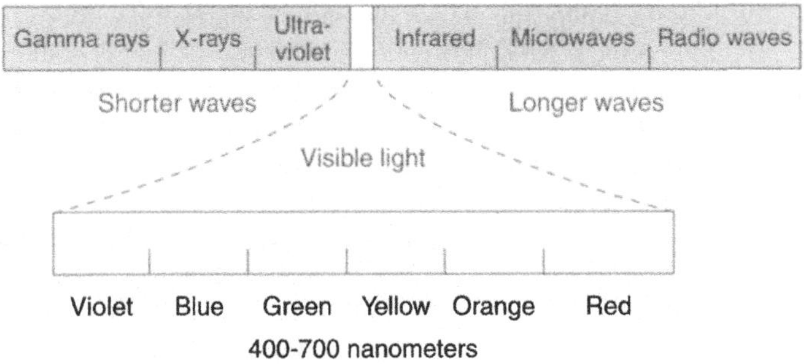

Figure 2-4

The electromagnetic spectrum. Visible light is only a small portion of the electromagnetic radiation that can be detected by various instruments.

Particle properties of light

Light is such a complicated phenomena that no one model can be devised to explain its nature. Although light is generally thought of as acting like an electric wave oscillating in space accompanied by an oscillating magnetic wave, it can also act like a particle. A "particle" of light is called a **photon**, or a discrete packet of electromagnetic energy.

Most visible objects are seen by reflected light. There are a few natural sources of light, such as the Sun, stars, and a flame; other sources are man-made, such as electrical lights. For an otherwise non-luminous object to be visible, light from a source is reflected off the object into our eye. The property of **reflection**, that light can be reflected from appropriate surfaces, can most easily be understood in terms of a particle property, in the same sense that a ball bounces off a surface. A common example of reflection is mirrors, and in particular, telescope mirrors that use curved surfaces to redirect light received over a large area into a smaller area for detection and recording.

When reflection occurs in particle-particle interactions (for example, colliding billiard balls), it's called **scattering** — light is scattered (reflected) off molecules and dust particles that have sizes comparable to the wavelengths of the radiation. As a consequence, light coming from an object seen behind dust is dimmer than it would be without the dust. This phenomena is termed **extinction**. Extinction can be seen in our own Sun when it becomes dimmer as its light passes through more of the dusty atmosphere as it sets. Similarly, stars seen from Earth seem fainter to the viewer than they would if there were no atmosphere. In addition, short wavelength blue light is preferentially scattered; thus objects look redder (astronomers refer to this as **reddening**); this occurs because the wavelength of blue light is very close to the size of the particles that cause the scattering. By analogy, consider ocean waves — a row boat whose length is close to the wavelength of the waves will bob up and down, whereas a long ocean liner will scarcely notice the waves. The Sun appears much redder at sunset. The light of stars also redden in passing through the atmosphere. You can see the scattered light by looking in directions away from the source of the light; hence the sky appears blue during the day.

Extinction and reddening of starlight are not caused by just the atmosphere. An exceedingly thin distribution of dust floats between the stars and affects the light that we receive as well. Astronomers must take into account the effect of dust on their observations to correctly describe the conditions of the objects that emit the light. Where interstellar dust is especially thick, no light passes through. Where dust clouds reflect starlight back in our direction, the observer may see blue interstellar wispiness like thin clouds surrounding some stars, or a **nebula** (to use the Latin word for cloud). A nebula formed by scattering of blue light is called a reflection nebulae.

Wave properties of light

Most properties of light related to astronomical use and effects have the same properties as waves. Using an analogy to water waves, any wave can be characterized by two related factors. The first is a **wavelength** (λ) the distance (in meters) between similar positions on successive cycles of the wave, for example the crest-to-crest distance. The second is a **frequency** (*f*) representing the number of cycles that move by a fixed point each second. The fundamental characteristic of a wave is that multiplication of its wavelength by its frequency results in the speed with which the wave moves forward. For electromagnetic radiation this is the speed of light, $c = 3 \times 10^8$ m/sec = 300,000 km/sec. The mid-range of visible light has a wavelength of $\lambda = 5500$ Å $= 5.5 \times 10^{-7}$ m, corresponding to a frequency f of 5.5×10^{14} cycles/sec.

When light passes from one medium to another (for example, from water to air; from air to glass to air; from warmer, less dense regions of air to cooler, denser regions and vice-versa) its direction of travel changes, a property termed **refraction**. The result is a visual distortion, as when a stick or an arm appears to "bend" when put into water. Refraction allowed nature to produce the lens of the eye to concentrate light passing through all parts of the pupil to be projected upon the retina. Refraction allows people to construct lenses to change the path of light in a desired fashion, for instance, to produce glasses to correct deficiencies in eyesight. And astronomers can build refracting telescopes to collect light over large surface areas, bringing it to a common focus. Refraction in the non-uniform atmosphere is responsible for mirages, atmospheric shimmering, and the twinkling of stars. Images of objects seen through the atmosphere are blurred, with the atmospheric blurring or astronomical "seeing" generally about one second of arc at good observatory sites. Refraction also means that positions of stars in the sky may change if the stars are observed close to the horizon.

Related to refraction is **dispersion**, the effect of producing colors when white light is refracted. Because the amount of refraction is wavelength dependent, the amount of bending of red light is different than the amount of bending of blue light; refracted white light is thus dispersed into its component colors, such as by the prisms used in the first spectrographs (instruments specifically designed to disperse light into its component colors). Dispersion of the light forms a **spectrum**, the pattern of intensity of light as a function of its wavelength, from which one can gain information about the physical nature of the source of light. On the other hand, dispersion of light in the atmosphere makes stars undesirably appear as little spectra near the horizon. Dispersion is also responsible for **chromatic aberration** in telescopes — light of different colors is not brought to the same focal point. If red light is properly focused, the blue will not be focused but will form a blue halo around a red image. To minimize chromatic aberration it is necessary to construct more costly multiple-element telescope lenses.

When two waves intersect and thus interact with each other, **interference** occurs. Using water waves as an analogy, two crests (high points on the waves) or two troughs (low points) at the same place **constructively interfere**, adding together to produce a higher crest and a lower trough. Where a crest of one wave, however, meets a trough of another wave, there is a mutual cancellation or **destructive interference**. Natural interference occurs in oil slicks, producing colored patterns as the constructive interference of one wavelength occurs where other wavelengths destructively interfere. Astronomers make use of interference as another means of dispersing white light into its component colors. A **transmission grating** consisting of many slits (like a picket fence, but numbering in the thousands per centimeter of distance across the grating) produces constructive interference of the various colors as a function of angle. A **reflection grating** using multiple reflecting surfaces can do the same thing with the advantage that all light can be used and most of light energy can be thrown into a specific constructive interference region. Because of this higher efficiency, all modern astronomical spectrographs use reflection gratings.

A number of specialized observing techniques result from application of these phenomena, of which the most important is **radio interferometry.** The digital radio signals from arrays of telescopes can be combined (using a computer) to produce high-resolution (down to 10^{-3} second of arc resolution) "pictures" of astronomical objects. This resolution is far better than that achievable by any optical telescope, and thus, radio astronomy has become a major component in modern astronomical observation.

Diffraction is the property of waves that makes them seem to bend around corners, which is most apparent with water waves. Light waves are also affected by diffraction, which causes shadow edges to not be perfectly sharp, but fuzzy. The edges of all objects viewed with waves (light or otherwise) are blurred by diffraction. For a point source of light, a telescope behaves as a circular opening through which light passes and therefore produces an intrinsic **diffraction pattern** that consists of a central disk and a series of fainter diffraction rings. The amount of blurring as measured by the width of this central diffraction disk depends inversely on the size of the instrument viewing the source of light. The pupil of the human eye, about an eighth of an inch in diameter, produces a blurring greater than one arc minute in angular size; in other words, the human eye cannot resolve features smaller than this. The Hubble Space Telescope, a 90-inch diameter instrument orbiting Earth above the atmosphere, has a diffraction disk of only 0.1 second of arc in diameter, allowing the achievement of well-resolved detail in distant celestial objects.

The physical cause of diffraction is the fact that light passing through one part of an opening will interfere with light passing through all other parts of the opening. This self-interference involves both constructive interference and destructive interference to produce the diffraction pattern.

Kirchoff's three types of spectra

Both dispersive and interference properties of light are used to produce spectra from which information about the nature of the light-emitting source can be gained. Over a century ago, the physicist Kirchoff recognized that three fundamental types of spectra (see Figure 2-5) are directly related to the circumstance that produces the light. These Kirchoff spectral types are comparable to Kepler's Laws in the sense that they are only a description of observable phenomena. Like Newton, who later was to mathematically explain the laws of Kepler, other researchers have since provided a sounder basis of theory to explain these readily observable spectral types.

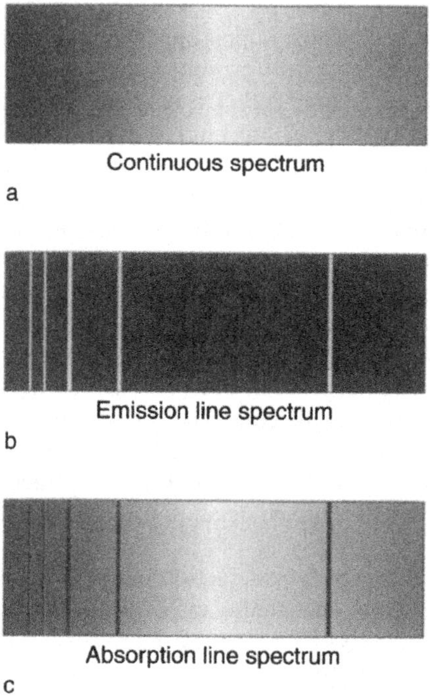

Figure 2-5

Kirchhoff's three types of spectra. a) A continuous spectrum (blackbody spectrum) is radiation produced by warm, dense material; b) an emission line spectrum (bright line spectrum) is radiation created by a cloud of thin gas; and c) an absorption line spectrum (dark line spectrum) results from light passing through a cloud of thin gas.

Kirchoff's first type of spectrum is a **continuous spectrum**: Energy is emitted at all wavelengths by a luminous solid, liquid, or very dense gas — a very simple type of spectrum with a peak at some wavelength and little energy represented at short wavelengths and at long wavelengths of radiation. Incandescent lights, glowing coals in a fireplace, and the element of an electric heater are familiar examples of materials that produce a continuous spectrum. Because this type of spectrum is emitted by any warm, dense material, it is also called a **thermal spectrum** or **thermal radiation**. Other terms used to describe this type of spectrum are **black body spectrum** (since, for technical reasons, a perfect continuous spectrum is emitted by a material that is also a perfect absorber of radiation) and **Planck radiation** (the physicist Max Planck successfully devised a theory to describe such a spectrum). All these terminologies refer to the same pattern of emission from a warm dense material. In astronomy, warm interplanetary or interstellar dust produces a continuous spectrum. The spectra of stars are roughly approximated by a continuous spectrum.

Kirchoff's second type of spectrum is emission of radiation at a few discrete wavelengths by a tenuous (thin) gas, also known as an **emission spectrum** or a **bright line spectrum**. In other words, if an emission spectrum is observed, the source of the radiation must be a tenuous gas. The vapor in fluorescent tube lighting produces emission lines. Gaseous nebulae in the vicinity of hot stars also produce emission spectra.

Kirchoff's third type of spectrum refers not to the source of light, but to what might happen to light on its way to the observer: The effect of a thin gas on white light is that it removes energy at a few discrete wavelengths, known as an **absorption spectrum** or a **dark line spectrum**. The direct observational consequence is that if absorption lines are seen in the light coming from some celestial object, this light must have passed through a thin gas. Absorption lines are seen in the spectrum of sunlight. The overall continuous spectrum nature of the solar spectrum implies that the radiation is produced in a dense region in the Sun, then the light passes through a thinner gaseous region (the outer atmosphere of the Sun) on its way to Earth. Sunlight reflected from other planets shows additional absorption lines that must be produced in the atmospheres of those planets.

Wien's and Stefan-Boltzman's Laws for Continuous Radiation

Kirchoff's three types of spectra give astronomers only a general idea of the state of the material that emits or affects the light. Other aspects of the spectra allow more of a quantitative definition of physical factors. Wien's Law says that in a continuous spectrum, the wavelength at which maximum energy is emitted is inversely proportional to temperature; that is, λ_{max} = constant / T = 2.898 × 10-3 K m / T where the temperature is measured in degrees Kelvin. Some examples of this are:

Object	*Temperature (T)* λ_{max}	*Spectral Region*
Person	300 K	9.7×10^{-6} m = 97,000 Å Infra-red
Sun	5,800 K	5.0×10^{-7} m = 5000 Visible
The star Vega	10,000 K	2.9×10^{-7} m = 2700 Ultraviolet

The **Stefan-Boltzman Law** (sometimes called Stefan's Law) states that the total energy emitted at all wavelengths per second per unit surface area is proportional to the fourth power of temperature, or energy per second per square meter = σT^4 = 5.67 × 10-8 watts / (m^2 K^4) T^4 (see Figure 2-6).

This simple principle produces a relationship between the total energy emitted by an object each second, the **luminosity** L, the radius r of a celestial object, assumed spherical, and the object's surface temperature. The total energy emitted per second = surface area × energy per second emitted by each unit area, or algebraically,

$$L = 4 \pi r^2 \sigma T^4$$

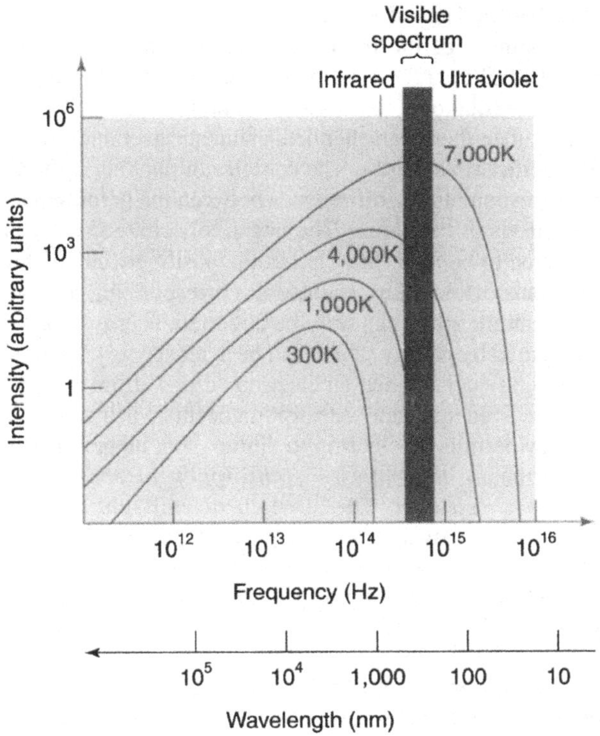

Figure 2-6

A graphical representation of continuous spectra for light sources of different temperature. Wien's Law is shown by the peak of radiation at shorter wavelengths for higher temperatures. The Stefan-Boltzman Law is shown by the larger areas under each curve (representing the total energy emission at all wavelengths) for higher temperatures.

Quantitative Analysis of Spectra

The development of the theory of quantum mechanics led to an understanding of the relationship between matter and its emission or absorption of radiation. If atoms are far enough apart that they do not affect each other, then each chemical element can emit or absorb light only at specific wavelengths. The energies of photons at these wavelengths correspond to the differences between the permitted energies of the electrons of the atoms. The negatively charged electrons can be considered to be in "orbits" about the positively charged protons in the nucleus of the atom, each orbit corresponding to a different energy. Quantum orbits can only be at certain energy levels, unlike orbits governed by gravity which can be at any energy. Emission of a photon of light occurs when an electron "drops" from a high energy state to a lower energy state. Absorption occurs when a photon *of the right energy* permits an electron to "jump" to a higher energy state. Most importantly, *the pattern of absorption or emission lines is unique to each element*. The strength of emission or absorption depends on how many atoms of the particular element are present as well as the temperature of the material, thus permitting both temperature and the chemical composition of the material producing the spectrum to be determined.

If atoms are progressively jammed closer together, the wavelengths of emission or absorption by any given atom will be slightly changed, thus some atoms will emit/absorb at slightly longer wavelengths and others will emit/absorb at slightly shorter wavelengths. The majority of atoms will emit/absorb at the same wavelengths that they would if unaffected by neighboring atoms. Astronomers therefore can differentiate between circumstances where the emitting or absorbing atoms are thinly dispersed (the spectral features will look very sharp) and where they are tightly packed together (the spectral features are broadened). In the extreme case of high density, the emission lines become completely blurred together and one observes a continuous spectrum.

Doppler's Law

If a light source and observer are approaching each other, the observed wavelength of any spectral feature is shorter than what would be measured if the two were at rest with respect to each other. On the other hand, if the two are moving apart, the observed wavelength appears longer that the wavelength that would be measured at rest. The **Doppler Shift** or **Doppler Effect** is a recognition that *the change of wavelength* ∅λ of a given spectral feature depends on the relative velocity of the source along the line of sight:

$$\Delta\lambda/\lambda_o = (\lambda-\lambda_o)/\lambda_o = v/c$$

Carol: Are the boxes highlighted below in MathType? If so, please create the code (I think the small circle on the second one should be a 0 not a circle with a dot like we've used before.). Thanks—Alissa

where Δ is the observed wavelength, Δ_0 is the **rest** or **laboratory wavelength**, v is the velocity toward (negative) or away (positive) from the observer, and c is the speed of light. Relative motion of a light source toward the observer results in a **blueshift** of the spectrum, as all wavelengths are measured shorter or bluer; relative motion of a light source away from the observer results in a **redshift**. This doesn't mean that the light literally turns blue or red, it means that the light has its color shifted toward the shorter or longer wavelength region of the spectrum, respectively (see Figure 2-7); in most situations the shift is very small because velocities are small compared to the speed of light. This simple form of the Doppler Law holds only if the velocity v is small with respect to the speed of light. A more complicated equation formulated by Einstein in his theory of relativity must be used if the source is moving near the speed of light.

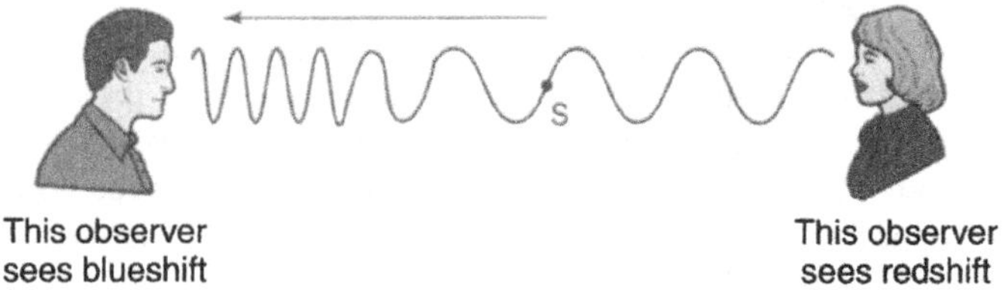

Figure 2-7

The Doppler effect. The frequency of light waves appears to change, varying with the relative velocity of the source (S) and the observer. If the source and observer are drawing closer together, the observed frequency is higher that the emitted frequency, and the observer sees blueshift. If the source and observer are getting farther apart, the observer sees redshift.

CHAPTER 3
OBSERVING THE SKY

Naked-Eye Astronomy

Human awareness of the universe began when people realized that they could observe objects in the sky, and that what they saw changed over days, months, and years.

The sky

To any observer on the ground looking at the sky with the naked eye, the sky appears to be a vast spherical bowl, a **celestial sphere** that extends from all points along the horizon to the **zenith,** the point directly overhead.

Astronomical objects seen in the sky are so far away that observers viewing them without the aid of a telescope have no intuitive sense of which objects are closer than others. This lack of depth perception causes everything to appear equidistant on the sky. Any object's position on this sphere can be determined by two coordinates, which designate the object's horizontal and vertical location. The vertical coordinate is determined by measuring an **altitude angle** upwards from the nearest point on the horizon, and the horizontal coordinate is established by measuring an **azimuth angle** from due north eastwards along the horizon to that nearest point. This system is called either the **horizon coordinate system** or the **alt-azimuth system** (see Figure 3-1).

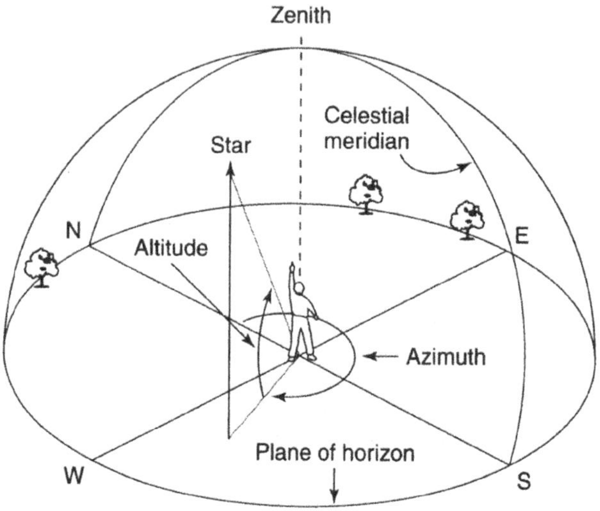

Figure 3-1

The horizon coordinate system.

Stars and constellations

About 6,000 stars are visible to the naked eye when you can achieve the darkest conditions. The positions of the stars relative to each other remain fixed from night to night and year to year. In general, stars in the same area of the sky have no physical relationship to each other, but the very human tendency to impose order upon otherwise random distributions yields patterns of brighter stars, or **constellations.** Many constellations of the Northern Hemisphere have been inherited from antiquity, including Ursa Major, the Big Bear, of which the more familiar Big Dipper is part, and Orion, the hunter, which can be seen in the winter sky. Many Southern Hemisphere constellations were defined in the last century to fill in unlabeled regions of the sky.

The sky is now officially divided into 88 constellations, which are used in modern astronomy for naming purposes. For example, the star alpha Ursa Majoris (α UMa) is located in the Big Bear constellation. Therefore, *Ursa Majoris* comes from the name of the constellation, Ursa Major, and the Greek letter α indicates that it's the brightest star within that constellation. In addition to modern names, some 90 or so stars also have names from antiquity. For instance, α UMa is also known as Dubhe.

Over the course of the night, stars move across the sky from east to west as a consequence of Earth's rotation on its axis. The stars appear to move in circular paths around a **celestial pole,** or either of two points on the celestial sphere where the extensions of Earth's axis would intersect. In the Northern Hemisphere, the celestial pole is coincidentally marked by the relatively bright star alpha Ursa Minoris, also known as Polaris. Simple geometry shows that the altitude angle of the pole star above the northern horizon is equivalent to the latitude of the observer on Earth.

To an observer in the Northern Hemisphere, stars that are always above the northern horizon are known as **circumpolar stars;** an observer in the Southern Hemisphere would see circumpolar stars around the south celestial pole. Stars that are further to the south and that rise and set sometime during the night are called **equatorial stars.** Equatorial stars rise in the east, move diagonally into the southern sky, achieving their highest position above the horizon on the **meridian** (the great circle that extends from due north on the horizon, through the zenith, to due south on the horizon). From the meridian, these stars move westward until they set below the western horizon.

The **celestial equator** is that great circle formed on the celestial sphere by extending the plane of the Earth's equator. The equator intersects the horizon due west and due east. An immediate consequence of Earth's rotation is identifying the preferred directions of north, south, east, and west around the horizon, called **cardinal directions,** by which humans naturally orient themselves.

Minute by minute, over the course of the night, both the altitude and azimuth of a star continually change. To more easily observe and track celestial objects, scientists have defined celestial coordinate systems that are fixed upon the sky and thus move with the stars. The equatorial coordinate system is a projection onto the sky of Earth's latitude and longitude coordinate system. **Celestial latitude,** known as **declination,** is the angular position north or south of celestial equation. **Celestial longitude,** measured around the celestial equation, is known as the right ascension (see Figure 3-2). Declination is measured in degrees, minutes of arc, and seconds of arc (see Chapter 2). Recognizing that the stars appear to move once around the sky in one day (24 hours), right ascension is measured not in degrees, but in *hours,* with 24 *hours of right ascension* (to distinguish from 24 *hours of time*) equal to 360 degrees.

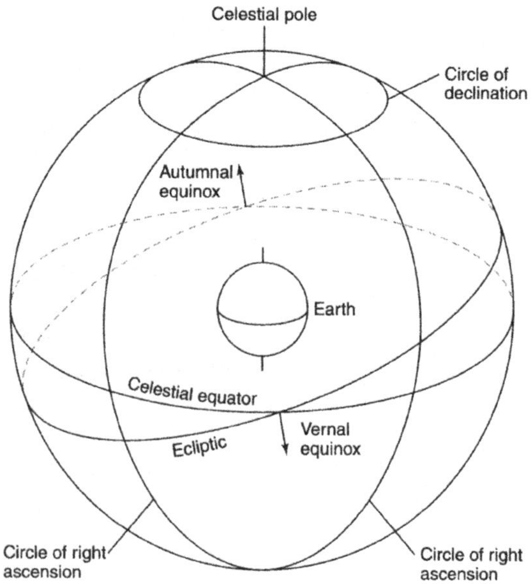

Figure 3-2

Celestial sphere features.

The Sun

Like the stars, the Sun rises and sets every day. However, unlike the fixed motion of the stars, the Sun's daily path across the sky varies throughout the year. Twice annually, at the **vernal equinox** (about March 21) and the **autumnal equinox** (about September 21), the Sun's position coincides with the celestial equator; it rises due east, moves across the sky following the path of the equator, and sets due west (see Figure 3-3). At this time of year, the length of the day is the same as the length of the night. During the summer, however, the solar position is north of the equator, achieving a maximum northernmost declination of +23.5 degrees at the time of the **summer solstice,** around June 21. In the winter, the solar position is reversed, with the Sun at its maximum southernmost declination of –23.5 degrees at the time of **winter solstice,** December 21. The terms equinox and solstice mark not only specific times of the year, but also specific points in the sky. The vernal equinox is that position on the celestial equator where the Sun crosses from the southern into the northern sky. Celestial longitude (right ascension) is measured eastward around the equator from this point.

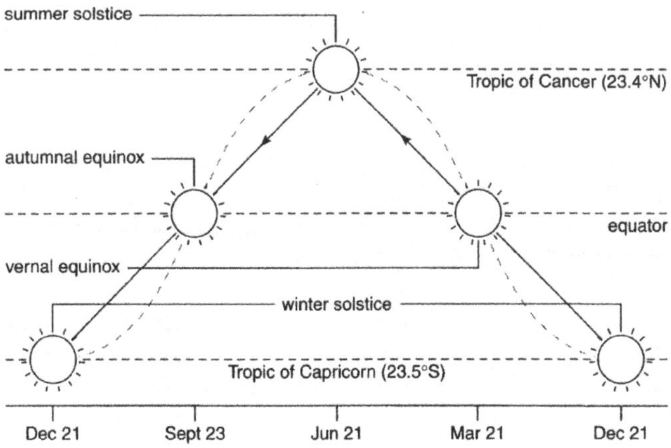

Figure 3-3

The Sun's seasonal positions in the sky.

Relative to the background stars, the Sun also moves about one degree eastward each day, thus encircling the sky in the course of a year. This **ecliptic,** or the great circle on the celestial sphere defined by the solar path, is tilted with respect to the celestial equator by 23.5 degrees. Throughout the year, the ecliptic passes through the **12 zodiacal constellations.** Because the apparent solar movement is caused by Earth's motion around the Sun, the **ecliptic plane** is also a projection of Earth's orbit about the Sun. That Earth's rotational equatorial plane does not perfectly match its **orbital plane** (the ecliptic plane) about the Sun is directly responsible for the annual north-south movement of the Sun and the consequent seasons.

Time

The length of the year is based on the time that the Sun takes to complete its annual path around the sky. During the year, the stars and constellations visible at night change as the seasons pass through their cycle. A year is, in fact, 365.25 days long. To keep the calendar in sync with the actual seasonal passage of time, every four years we have a leap year, or a year of 366 days. The additional day, February 29, makes up for the time lost annually when the 365.25 day cycle is computed as 365 days.

The length of a day is based on the time it takes for the Sun to complete its daily path across the sky, from its crossing of the meridian at noon to its next crossing of the meridian the next day. Daily time based on the hourly position of the Sun in the sky, relative to its noontime position each day, is called **apparent solar time.**

To astronomers studying the stars, however, the time of day alone is not a sufficient method for determining which stars can be observed at any given time. Astronomers need to know both the time of day and where the Sun is in the sky to find out which stars they can observe. More specifically, what is of interest to an astronomer is the **sidereal time,** or the right ascension of the stars that are crossing the meridian at any moment. Sidereal time and solar time are not equal because of the annual motion of the Sun across the sky. The true rotation period of Earth is 23 hours, 56 minutes, but the length of the day is four minutes longer. In one rotation of Earth, the Sun has moved roughly one degree across the sky, thus Earth has to rotate a little bit longer to get the Sun back to the same position on the sky. The solar day is 24 hours long.

The actual length of any given day varies throughout the course of the year for two reasons. First, Earth's orbit around the Sun is not a circle, but an ellipse. The Sun's apparent motion across the sky therefore varies slightly, moving more than the average one degree per day in January, but a bit less than the average in July. Second, the solar motion along the ecliptic moves parallel to the celestial equator at the times of the solstices when the Sun is farthest north or south of the equator, but at the times of equinoxes, its motion is at an angle to the celestial equator. What is important for the length of the day is only that part of the solar motion parallel to the equator. This produces a second variation to the length of the day.

The difference between apparent solar time (which would be shown by a sundial) and mean solar time (which would be shown by a clock or watch) is called the **equation of time.** This is a complicated effect involving orbital mechanics (the motion around an ellipse is not uniform), perspective (a moving object closer to the observer has a greater apparent motion), the tilt of the ecliptic to the equator (hence the daily component of solar motion across the sky varies from equinox to solstice), and a geometrical factor resulting from the fact that we are dealing with a spherical coordinate system; this latter you can demonstrate to yourself by placing your thumb on the equator of a globe. Note how much longitude angle it covers. Now place your thumb near the north pole. Your thumb will cover the same area, but as measured in longitude, it covers a much bigger longitudinal angle. And it's that angle that is important for rotation and hence the length of a day. Noon, as defined by mean solar time, can occur as much as 16 minutes late in October or 14 minutes early in February compared to the Sun's actual crossing of the meridian.

The Moon

Observation of the Moon shows that it not only appears to change its position by moving around Earth, but at the same time, the fraction of its surface that is illuminated by sunlight (its **phase**) also changes (see Figure 3-4). When the Moon is directly opposite the position of the Sun, it appears totally illuminated, or in **full phase.** When the Moon is viewed at an angle of 90 degrees from the position of the Sun, its surface appears half illuminated, or in its **quarter phase.** Between full and quarter Moon, the phase is said to be **gibbous.** If less than half the Moon is illuminated by sunlight, the phase is **crescent.** When the Moon is in the direction of the Sun and the side toward Earth is its dark or shadowed half, the Moon is said to be **new.** The cycle from **new Moon waxing** (increasing illumination) to first quarter to full Moon, then **waning** (decreasing illumination) to third quarter and back to new Moon takes 29.5 days, a period adopted as the basis of our calendrical month. The lunar phases are observed because of the changing geometrical orientation between the Sun, Moon, and Earth over the course of the lunar month, *not* the result of Earth casting its shadow on the Moon.

The Moon takes 27.3 days to move once (360 degrees) around Earth as determined by its **orbital position** relative to the stars; the **orbital period** is therefore also an *example* of a **sidereal period,** a period measured relative to the stars. This 27.3 days can also be called a **sidereal month.** The time from full Moon to full Moon is longer, 29.5 days, because two motions are involved: the motion of the Moon around Earth and the motion of Earth around the Sun. To become full again, the Moon must move more than 360 degrees around Earth. The lunar month, or **lunation,** is thus an *example* of a **synodic period,** or a period produced as the consequence of *two* motions. The term **synodic month** is also applied to this 29.5 day period. Another example of the difference between a sidereal period and a synodic period is the rotation of Earth (23 hours, 56 minutes) and the length of a day (24 hours).

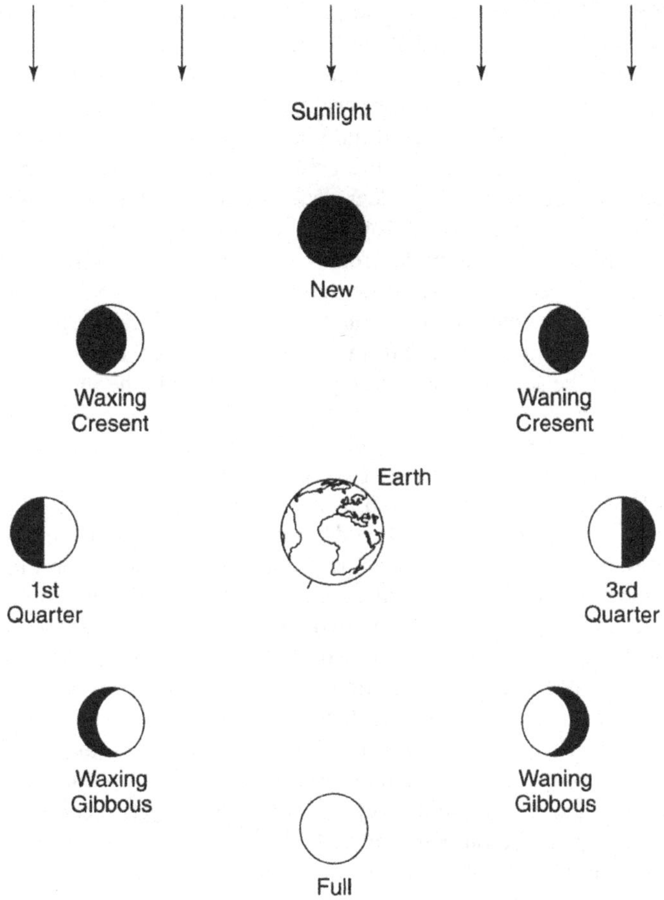

Figure 3-4

Phases of the Moon.

The actual path across the sky of the Moon is quite complicated. The lunar orbital plane is tilted about 5 degrees 9 minutes with respect to ecliptic plane. The orientation of these two planes rotate with respect to each other over an 18.6 year period. In consequence, relative to the celestial equator, the lunar orbital plane has a variable tilt that periodically oscillates from 23.5 degrees minus 5 degrees 9 minutes equals 18.3 degrees, to 23.5 degrees plus 5 degrees 9 minutes equals 28.7 degrees. In other words, over the course of the month, the Moon can oscillate north and south across the celestial equator between declinations of as little as $\pm$18.3 degrees to as much as $\pm$28.7 degrees. Each month the path of the Moon across the sky differs slightly from its path the prior month and from its path the next month. In contrast, each year the solar path around the sky is essentially the same.

Astronomers find particularly interesting the two intersection points, or **nodes,** between the great circles of the ecliptic plane and the lunar orbit. If the Sun and Moon in their motions about Earth move simultaneously through nodal positions, an eclipse occurs (see Figure 3-5). If the Moon and Sun are at the same node, the Moon passes in front of the Sun and a **solar eclipse** results, lasting only a few minutes in duration. If the two are at opposite nodes, then the Moon passes through the shadow of the Earth and a longer duration **lunar eclipse** occurs. The cycle of occurrence of eclipses is quite complicated because three factors are involved. The Sun moves around the sky in 365.25 days. The Moon moves once around the sky every 27.3 days. The direction to a node (where the two orbital planes intersect) moves around the sky in 18.6 years. To get an eclipse, all three positions (Moon, Sun, and node) have to be along a straight line through Earth. This event does not happen with a simple pattern of regular time intervals.

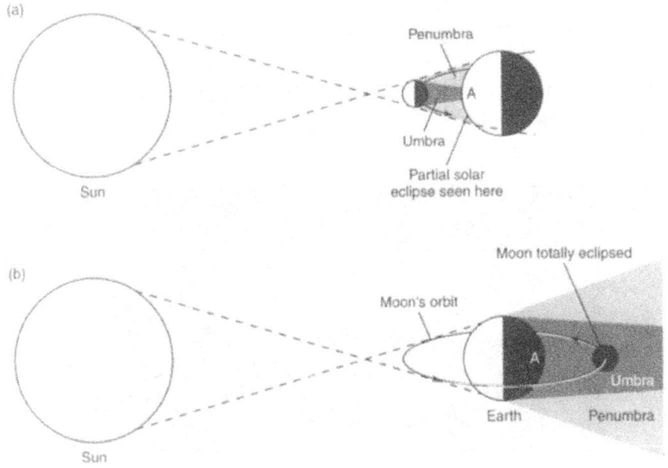

Figure 3-5

Solar and lunar eclipses.

Planets

Naked-eye observers in antiquity recognized five other objects that move across the sky relative to the stars — the planets. The word *planet* comes from the Greek, meaning "wanderer." Planetary motion is more complicated than that of the Moon or Sun, which move systematically west to east relative to the stars. The planets' usual motion is west to east, termed **direct** or **prograde motion.** But in each synodic period, planets show a brief period of motion in the **opposite** or **retrograde direction.**

Retrograde motion is the result of perspective. Consider, for example, driving on a freeway. You look forward toward a slower moving car. Relative to the distant mountains, it is clear that this car is moving in the same direction you are moving. But as you catch up to and bypass the slower car, you see it appear to be moving backwards compared to the distant horizon. As you pull ahead, once again looking back at the slower car shows it to be moving forward. Observation of other planets from Earth produces the same apparent reversal of motion relative to the background of stars. (See Figure 3-6.)

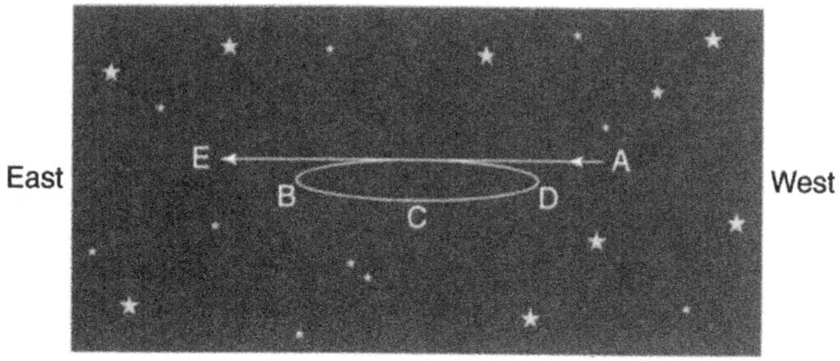

Figure 3-6

Retrograde motion. A planet displaying retrograde motion appears to briefly move east to west rather than west to east.

Telescopes and Observatories

The primary purpose of a **telescope** is to collect light over a large surface area and secondarily to produce a magnification of the image of the objects under study. The site at which one or more telescopes have been constructed is known as an **observatory.**

Refracting telescopes

A **refracting telescope,** or **refractor,** uses the property of refraction (see Chapter 2) to bend the paths of light entering at every position on the **objective lens** to bring it to a common focus (see Figure 3-7). The basic function is similar to that of the eye of a camera. The largest existing refractor is the 1.02-meter-diameter Yerkes telescope in

Wisconsin. Refractors have gone out of favor for astronomical work due to the enormous size of the glass lenses (the Yerkes objective lens weighs 2 tons), the bending of the long tube of the telescope due to its weight, their awkwardness because of the length of the tube, and the cost of the large enclosure needed to shelter such a telescope. Also, as objective lenses are made larger, they become thicker, causing progressively significant loss of light by absorption in the glass.

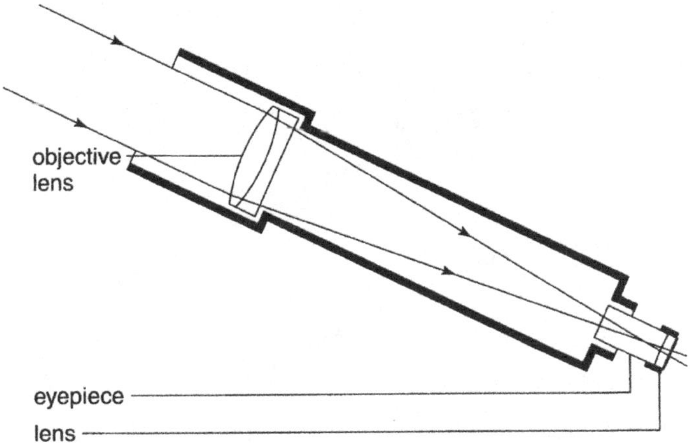

Figure 3-7

A refracting telescope.

Reflecting telescopes

A **reflecting telescope,** or **reflector,** uses a curved mirror as its light gathering surface (**primary mirror**) and other mirrors and optical elements to bring all incident light to a common focus (see Figure 3-8). Modern reflectors have a number of advantages over refractors in that they can be constructed with larger light gathering surfaces (they can be structurally supported on the back of the mirror), are mechanically more stable, are physically smaller instruments (multiple reflections allow the light path to be "folded over"), require a smaller and less expensive observatory building, and are much easier to use. These reflectors do have the disadvantage that they require more critical alignment of the optical elements (the mirrors); but with modern computers and laser technology, alignment can be easily corrected. The most recent large telescopes are also able, under computer control, to modify the surface configuration of their primary mirrors (**active optics**) to correct for the visually blurring effects of the atmosphere.

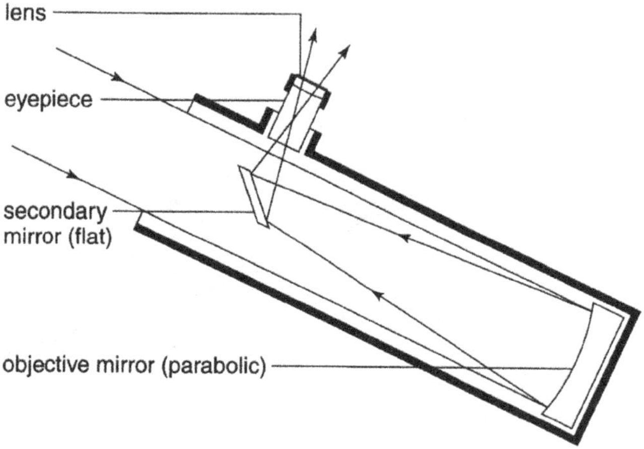

Figure 3-8

A reflecting telescope.

Detectors

Although the original **detector** of light was the human eye, astronomers now rely upon a wide variety technological devices for the recording of electromagnetic radiation. These devices are not only far more sensitive than the eye (a fully dark-adapted eye detects only

2 to 3 percent of the incident photons, whereas a modern electronic light detector records 90 percent or more of the photons), but often can add (integrate) the energy received over long periods of time and in wavelength regions other than visible light.

The first device to replace the human eye for astronomical observation was the photographic plate. Modern devices, such as the **charge-coupled device** (CCD), are far more efficient than photography (only 1 to 5 percent of the light is actually absorbed to produce a photographic image). A CCD is an electronic device using silicon chips to absorb the photons received over a vast array of picture elements (pixels) to produce free electrons that can then be counted to form a digital picture, which can be manipulated by a computer program. (Commercial digital cameras use the same technology.) **Photometers** are used where the intensity of light is to be measured. Other specialized devices, often carried by artificial satellites, have been developed to explore other regions of the spectrum, from extremely short gamma- and X-radiation to long wavelength radio radiation.

CHAPTER 4
THE SOLAR SYSTEM

The solar system consists of the Sun, nine planets, some 60 or so moons, and assorted minor materials (asteroids, meteoroids, comets, dust, and gas). All of these objects are tiny in comparison to the distances that separate them. Imagine the solar system scaled down such that distances to the planets could be spaced along a 10-kilometer hiking trail. On such a scale the Sun would be represented by a ball only 2.3 meters in diameter. The distribution along the trail and the model sizes for each of the planets are shown in the following table:

Table 4-1: Scale Model of the Solar System

Object	***Position along the Trail***	***Diameter of Model Sphere***	***Comment***
Sun	At start of trail	2.3 meters	7.6 feet
Mercury	100 meters	0.8 cm	Small marble
Venus	180 meters	2.0 cm	Large marble
Earth	250 meters	2.2 cm	Large marble
Mars	390 meters	1.1 cm	Small marble
Jupiter	1.3 kilometers	24 cm	Small (youth) soccer ball
Saturn	2.4 kilometers	20 cm without rings	Small (youth) soccer ball
Uranus	4.9 kilometers	8.6 cm	Baseball
Neptune	7.6 kilometers	8.3 cm	Baseball
Pluto	10 kilometers, trail's end	0.2 cm	BB, or small ball bearing
Comets	To 25,000 kilometers	20 μ – 200 μ	Dust particles

On this scale, the nearest star is about 2.3 meters in diameter and about 70,000 kilometers distant from the Sun.

Origin and Evolution of the Solar System

The solar system was formed $4.6 \pm 0.1 \times 10^9$ years ago. Astronomers have recognized a number of observable facts about the solar system that are not otherwise the result of obvious physical laws (for example, Kepler's Laws of Planetary Motion, which are the direct result of the nature of gravity). But the foundation of science assumes that every observable property must result from some cause. These features must therefore be the direct result of how the solar system formed. The following list outlines these observable facts:

- All planetary orbits lie nearly in a single plane; in other words, the solar system is flat (the orbit of Pluto is an exception).
- The Sun's rotational equator is in the same flat plane.
- Planetary orbits are nearly circular (exceptions are Mercury and Pluto).
- The planets and Sun all revolve in the same direction, that is, a motion that is west to east across the sky as viewed from Earth (what astronomers refer to as *direct motion*).
- The Sun and planets all rotate in the same direction with **obliquities** (the tilt between the equatorial and orbital planes) generally small (exceptions are Venus, Uranus, and Pluto).
- Planets and most asteroids have similar rotational periods (exceptions are Mercury, Pluto, and Venus).
- Planets are regularly spaced (this is often expressed in the form of a simple mathematical progression, known as **Bode's law**).

- The major moons in planetary satellite systems resemble the solar system on a smaller scale (circular orbits, uniform direction of revolution, in a flat plane with regular spacing).
- Most angular momentum (≈ mass × velocity × orbital radius) of the solar system is in the planets (99.8%), whereas most of the mass of the solar system is in the Sun (99.8%). This may be expressed alternatively as a question: Why does the sun rotate so slowly?
- Differences in chemical composition exist throughout the solar system, with dense, metal-rich (terrestrial) planets found close to the sun, but giant, hydrogen-rich (gas) planets only in the outer part of the solar system. In addition, the chemical composition of meteorites, while similar, is not identical to all known planetary and lunar rocks.
- Comets exist in a much larger, spherical cloud surrounding the solar system.

Throughout the years, people have come up with a variety of theories to explain the observable features of the solar system. Some of these theories include so-called *catastrophe theories,* such as a near collision of the Sun with another star. Modern theory of planetary origins also explicitly rejects any idea that our solar system is unique or special, thus ruling out catastrophe theories. The **solar nebula theory** (also known as the **planetesimal hypothesis,** or **condensation theory**) describes the solar system as the natural result of the operation of the various laws of physics. According to this theory, before the planets and Sun were formed, the material that would become the solar system existed as part of a large, diffuse cloud of interstellar gas and dust (a **nebula**) composed primarily of hydrogen and helium with traces (2 percent) of other, heavier elements. Such clouds can be stable for very long periods of time with simple gas pressure (pushing outward) balancing the inward pull of the self-gravity of the cloud. But British theoretician James Jeans showed that the smallest disturbance (perhaps an initial compression begun by a shock wave from a nearby stellar explosion) allows gravity to win the competition, and gravitational contraction begins. The fundamental inability for gas pressure to permanently balance against self-gravity is known as the **Jeans Instability.** (An analogy would be a yardstick balanced on one end; the slightest displacement upsets the balances of forces and gravity causes the yardstick to fall over.)

During the nebula's gravitation collapse (**Helmholtz contraction**), gravity accelerated particles inward. As each particle accelerated, the temperature rose. If no other effect were involved, the temperature rise would have increased pressure until gravity was balanced and the contraction ended. Instead the gas particles collided with each other, with those collisions converting kinetic energy (the energy of a body that is associated with its motion) into an internal energy that atoms can radiate away (in other words, a cooling mechanism). About half the gravitational energy was radiated away, and half went into heating the contracting cloud; thus, gas pressure stayed below what was needed to achieve balance against the inward pull of gravity. As a result, the contraction of the cloud continued. The contraction occurred more quickly in the center, and the density of the center mass rose much faster than the density of the outer part of the nebula. When the central temperature and density became great enough, thermonuclear reactions began to provide significant energy—in fact, enough energy to allow the central temperature to reach the point where the resulting gas pressure could again supply balance against gravitation. The central region of the nebula becomes a new Sun.

A major factor in the formation of the Sun was **angular momentum,** or the momentum characteristic of a rotating object. Angular momentum is the product of linear momentum and the perpendicular distance from the origin of coordinates to the path of the object (≈ mass × radius × rotational velocity). In the same manner that a spinning skater revolves faster when her arms are pulled inward, the conservation of angular momentum causes a contracting star to increase in rotational velocity as the radius is reduced. As its mass shrank in size, the Sun's rotational velocity grew.

In the absence of other factors, the new Sun would have continued rapidly rotating, but two possible mechanisms slowed this rotation significantly. One was the existence of a **magnetic field.** Weak magnetic fields are present in space. A magnetic field tends to lock into material (think of how iron filings sprinkled onto a sheet of paper on top of a magnet line up, mapping out the pattern of magnetic field lines). Originally the field lines would have penetrated the stationary material of the nebula, but after it contracted, the field lines would have been rapidly rotating at the central Sun, but very slowly rotating in the outer part of the nebula. By magnetically connecting the inner region to the outer region, the magnetic field sped up the movement of the outer material, but slowed the rotation (**magnetic braking**) of the central solar material. Thus momentum was transferred outward to the nebular material, some of which was lost to the solar system. The second factor to slow the early Sun's rotation was most likely a powerful solar wind, which also carried away substantial rotational energy and angular momentum, again slowing the solar rotation.

Beyond the center of the nebula, angular momentum also played a significant role in the formation of the other parts of the solar system. In the absence of outside forces, angular momentum is conserved; hence, as the radius of the cloud decreased, its rotation increased. Ultimately, rotational motions balanced gravity in an equatorial plane. Above and below this plane, there was nothing to hold up the material, and it continued to fall into the plane; the **solar nebula** exterior to the new central Sun thus flattened into a rotating disk (see Figure 4-1). At this stage, the material was still gaseous, with lots of collisions occurring between the particles. Those particles in elliptical orbits had more collisions, with the net result being that all material was forced into more or less circular orbits, causing a rotating disk to be formed. No longer significantly contracting, the material of this protoplanetary disk cooled, but heating from the center by the new Sun resulted in a temperature gradient ranging from a temperature of approximately 2,000 K at the center of the nebula to a temperature of approximately 10 K at the edge of the nebula.

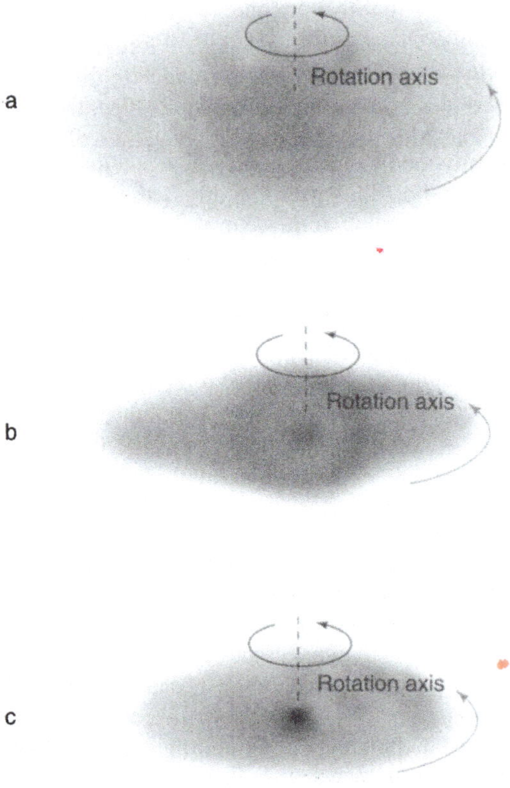

Figure 4-1

Collapse of interstellar cloud into star and protoplanetary disk.

Temperature affected which materials condensed from the gas stage to the particle (**grain**) stage in the nebulae. Above 2,000 K, all elements existed in a gaseous phase; but below 1,400 K, relatively common iron and nickel began to condense into solid form. Below 1,300 K, silicates (various chemical combinations with SiO_4) started to form. At much lower temperatures, below 300 K, the most common elements, hydrogen, nitrogen, carbon, and oxygen, formed ices of H_2O, NH_3, CH_4, and CO_2. Carbonaceous chondrites (with chondrules, or spherical grains that never were melted in later events) are the direct evidence that grain formation took place in the early solar system, with a subsequent amalgamation of these small solid particles into larger and larger objects.

Given the range of temperature in the **protoplanetary nebula,** only heavy elements were able to condense in the inner solar system; whereas both heavy elements and the much more abundant ices condensed in the outer solar system. Gases that didn't condense into grains were swept outward by radiation pressure and the stellar wind of the new Sun.

In the inner solar system, heavy element grains slowly grew in size, successively combining into larger objects (small moon-sized planets, or **planetesimals**). In the final stage, planetesimals merged to form the small handful of terrestrial planets. That smaller objects were present before the planets is shown by the leftover asteroids (too far from either Mars or Jupiter to become part of those surviving planets) and the evidence of impact cratering on the ancient surfaces of the large bodies that exist today. Detailed computations show that the formation of larger bodies in this manner produces final objects rotating in the same sense of direction as their motion about the Sun and with appropriate rotational periods. The condensation into a few objects orbiting the Sun occurred in more or less regularly spaced radial zones or annuli, with one surviving planet in each region.

In the outer solar system, **protoplanets** formed in the same manner as those in the inner solar system, but with two differences. First, more mass was present in the form of icy condensates; and second, the amalgamation of solid materials occurred in a region rich in hydrogen and helium gas. The gravitation of each growing planet would have affected the surrounding gas dynamics until gravothermal collapse occurred, or a sudden collapse of surrounding gas upon the rocky-icy protoplanets, thus forming the final nature of the gas giants. In the vicinity of the largest developing gas giants, the new planet's gravity affected the motions of surrounding, smaller objects with the evolution there being like a smaller version of the whole solar system. Thus, satellite systems ended up looking like the whole solar system in miniature.

Terrestrial Planets and Gas-Giant Planets

The goal of planetary astronomy is to understand both the differences and the similarities (called **comparative planetology**) of the major objects in the solar system, including their atmospheres, surfaces, internal structures, and other factors, such as magnetic fields. The study is also aimed at understanding the sequence of events and dates that define the history of the solar system. When studying the planets of the solar system (see Figure 4-2), the first thing one might notice is that the characteristics of the planets suggest a division into two fundamental types: the terrestrial (Earth-like) and the gas-giant (Jupiter-like) planets. These characteristics are compared in Table 4-2.

Table 4-2: Comparison of Planetary Properties

Property	***Terrestrial Planets***	***Gas-Giant Planets***
Location in solar system	Nearest Sun	Outer part of solar system
Size (or radius)	Small	Large
Mass	Non-massive	Massive
Density	Dense	Not dense
Structure	Solid, with iron cores	Mostly gaseous with solid cores

(continued)

Table 4-2 *(continued)*

Property	*Terrestrial Planets*	*Gas-Giant Planets*
Chemical composition	Heavy elements	Primarily hydrogen and helium
Atmosphere	Thin	Deep and thick
Exterior temperature	Warm	Cold
Rotation	Slow	Fast
Satellites	Few	Many
Ring systems	No	Yes

The outermost planet, Pluto, however, does not fit into either of these classifications: for example, it is small (really too small to be a terrestrial planet), but in the outer part of the solar system.

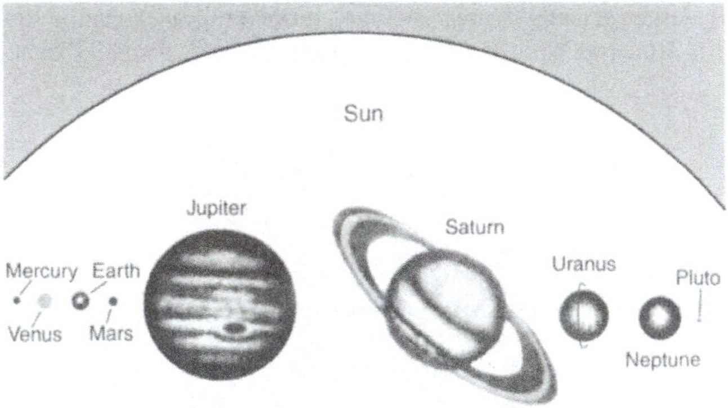

Figure 4-2

The planets of the solar system, drawn to scale.

Comparative Planetology — Terrestrial Planets

While each of the terrestrial planets (the Moon can be considered one of these) has its individual characteristics, many of their differences may be understood in the context of their cooling history: Smaller objects cool more quickly than large objects. The smallest terrestrial objects, the Moon and Mercury, froze solid relatively quickly. As a result, their surfaces date to the time of this formation and preserved on these solid surfaces is a history of events dating from very early times in the solar system. On the other hand, the largest terrestrial planet, Earth, is still in the process of cooling, with heat flowing outward from the hot core, driving convection in the mantle and producing the plate tectonic phenomena at the surface.

All the terrestrial planets have a similar crust-mantle-core structure (see Figure 4-3). There is a general trend for density to increase with the size of the object (even when corrected for compressional effects), indicating that a greater proportion of iron and nickel is present in progressively larger planetary cores. Mercury, however, is an exception to this rule; its iron core is far larger than expected for such a small planet.

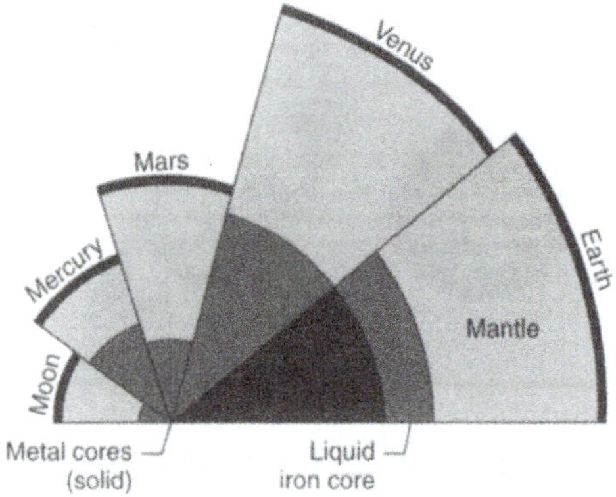

Figure 4-3

Schematic cross section of the terrestrial planets.

Comparative Planetology — Gas-Giant Planets

Although the four gas-giant planets are basically balls of hydrogen and helium gas and differ primarily only in mass, they have vastly different appearances. The progressive change of appearance in these planets, from the spectacular orange-reddish banding and belting of Jupiter to the deep blue, nearly featureless appearance of Neptune, may be attributed to a single factor: their outer temperature. This temperature results from the balance between thermal radiation of the planet versus the absorption of solar energy. These outer planets also have differences in their overall makeup, due to differences in their net chemical composition and to the manner in which the various chemical elements can exist at the temperatures and pressures found in the planetary interiors (see Figure 4-4).

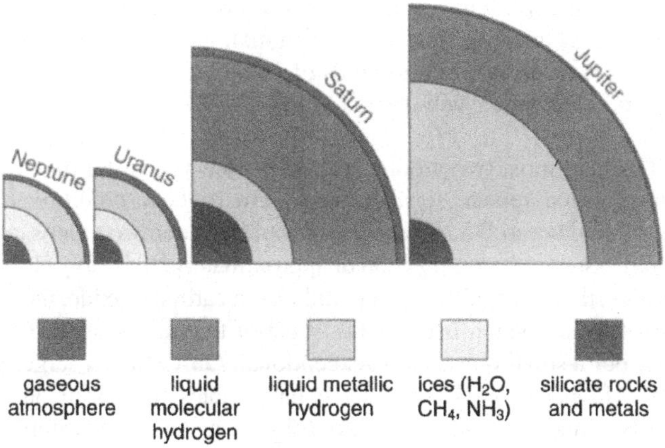

Figure 4-4

Comparison of the internal structure of the gas-giant planets.

Moons

The approximately 60 moons in our solar system are found primarily in orbit about the gas-giant planets. Because of the proximity of objects to each other and the relatively short time scales for gravitational modification of orbits, the lunar systems show many simple numerical relationships between their orbital periods (what astronomers term *resonances*). Ignoring the smallest objects, which appear to be debris from the collisional breakup of asteroids that has been captured into orbit after the formation of the planets, the moons are a distinct class of solar system object, chemically differentiated from both types of planets as well as other classes of objects in the solar system.

The four large moons of Jupiter, the so-called **Galilean moons** Io, Europa, Callisto, and Ganymede, probably formed in association with the formation of Jupiter itself; but the remaining 12 smaller satellites are probably captured asteroids. These four major moons are in almost perfect **gravitational resonance** with each other. Over the history of the solar system, their mutual gravitational pulls have produced respective orbital periods of 1.769 days, 3.551 days, 7.155 days, and 16.69 days, with period ratios of 1.00:2.00:2.02:2.33.

The innermost two moons are rocky objects like Earth's Moon, though Europa appears to have an icy crust, which could overlie a deeper liquid ocean. The lower densities of the outer two moons (about 2.0 g/cm^3) suggest a composition of approximately half heavy elements (iron and silicates) and half **ices** (solid water, carbon dioxide, methane, and ammonia), which is typical of most of the moons about the gas giants. For a small object, Io is exceptional. Only slightly larger than Earth's Moon, it would be expected to have cooled and frozen long ago, but it is actually the most volcanic object in the solar system. The source of energy that keeps its interior molten is the changing gravitational tides produced by Europa as Io sweeps by on its inner orbit every three and a half days. The gases released from volcanoes on Io have produced a donut-like belt of tenuous sulfur and sodium atoms about Jupiter. There also is evidence of ancient surface activity on Ganymede, which suggests that it too may have experienced some tidal heatings. Callisto, on the other hand, may have solidified so quickly that its heavier elements could not sink into the interior to form a core denser than the mantle.

Saturn has the largest family of moons whose compositions are again various combinations of rocky material and ice and whose orbits show many resonance relationships. These relationships include period-period resonances between moons in different orbits and also 1:1 resonances, where a smaller object may be trapped 60 degrees ahead or behind in the orbit of a larger object. For example, the small moons Telesto (25 km diameter) and Calypso (25 km) are trapped by Tethys (1048 km) in its orbits. Janus and Epimetheus share nearly the same orbit, switching places every time the inner one catches up to the outer one.

Saturn's large moon, Titan, has the densest atmosphere (mostly nitrogen with some methane and hydrogen) of any satellite. With a surface pressure about 40 percent that of Earth, this produces a greenhouse effect temperature of 150 K — about twice the expected value based only on absorption of sunlight.

Orbiting Uranus are four largish (radii 580–760 km) and one intermediate size (radius 235 km) moons, with about ten known smaller objects. This lunar family includes Miranda, probably the most bizarre object among all solar system satellites. Its surface shows evidence of past cataclysmic events (was it broken up in a collision and reassembled?), and possibly it is in the process of readjusting to an equilibrium structure as lighter ices rise and heavier materials sink. Contrary to expectation, the planet's moons do not show resonances between their orbital periods.

Neptune's lunar system is unusual in that its largest moon, Triton, is in a retrograde orbit tilted 23 degrees with respect to the planet's equator, and a second moon, Nereid, is in a very elongated orbit. Tidal stresses imposed on Triton by Neptune have caused internal heating and alteration of its icy surface, eliminating ancient craters. Its surface appears unique in that activity there is in the form of geysers — at a surface temperature of 37 K, absorption of sunlight vaporizes frozen nitrogen below the surface, which escapes by forcing itself through the overlying ices. Because the Moon orbits in a direction opposite to the rotation of the planet, tidal effects also are decelerating its motion, causing it to slowly spiral in toward the planet. Triton will move within Neptune's Roche Limit in perhaps 100 million years and be destroyed, and its material will be dispersed in a Saturn-like ring system. This suggests that Triton possibly was captured relatively recently, originally into an elliptical orbit that has been circularized by tidal effects.

Rings

All four of the outer planets in our solar system have rings composed of particles as small as dust to boulder-size materials orbiting in their equatorial planes. Jupiter is encircled by a tenuous ring of silicate dust, probably originating from particles chipped off the inner moons by the impact of micrometeorites. Uranus is orbited by 11 optically invisible, thin rings composed of boulder-size, dark particles; and Neptune has three thin and two broad rings, also composed of dark particles. The particles in the thin rings are unable to disperse due to the presence of **shepherd moons,** pairs of small moons only a few kilometers in diameter orbiting near the inner and outer edges of the rings. The shepherd moons' gravitational action confines small particles into a narrow ring at an intermediate orbital radius. The ring particles of Uranus and Neptune are dark because they are covered with dark organic compounds produced by chemical reactions involving methane.

It is Saturn that possesses the most extensive and obvious ring system, some 274,000 kilometers in diameter (see Figure 4-5). As seen from Earth, there is an apparent inner ring that extends inward to the top of the planet's atmosphere. Exterior to a large gap is a faint (or crape) ring, then a middle bright ring with a thin gap, the prominent Cassini's Gap, and finally an outer ring, Enke's Gap. Both the pattern of circular velocities as well as Earth-based radar studies show that the rings are composed of myriads of small particles, each orbiting as a tiny moon. These are highly reflective icy particles, from a few centimeters in size to a few meters in size.

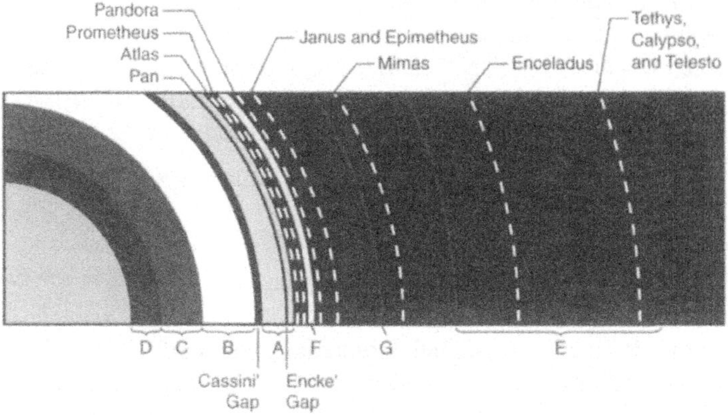

Figure 4-5

Saturn's ring system.

The rings of all the outer planets lie within each planet's **Roche limit,** the radial distance interior to which materials cannot coalesce into a single object under their own gravitation. In other words, the contrary gravitational pull on particles by the opposite sides of the planet is greater than the self-gravity between particles. If a satellite were to pass closer to the planet than the Roche limit (about 2.4 planetary diameters, depending upon the size, density, and structural strength of the satellite), it would be broken apart by the gravitational forces of the planet (another example of which are tidal forces).

The ring system of Saturn further illustrates the great variety of dynamical phenomena that are the result of gravitational attraction between systems of particles of greatly differing masses. First, the planet has an equatorial bulge; the slight excess of mass about the equator gravitationally perturbs the orbits of smaller objects (from dust particles to moons) into its equatorial plane; hence the ring system is flat. Most of the gaps in the rings (small particles) are due to orbital resonances with the larger satellites. For example, the moon Mimas produces Cassini's Gap where particles otherwise would be orbiting the planet with half that moon's orbital period. Enke's Gap, however, is the result of a clearing of particles by a small moon that orbits at that distance from the planet. That Saturn's ring system is composed of thousands of such rings also suggests that there are numerous shepherd moons, only a few of which have been discovered.

Minor Objects in the Solar System

Four basic categories of smaller materials exist in the solar system: meteoroids; asteroids (or minor planets); comets; and dust and gas. These categories are differentiated on the basis of chemistry, orbital characteristics, and their origins.

Meteoroids

Meteoroids are basically the smaller bodies in between the planets, defined as any rocky-metallic objects less than 100 meters, or alternatively 1 kilometer, in size. It is these objects that generally fall to Earth. While heated to incandescence by atmospheric friction during their passage through the atmosphere, they are termed **meteors**. A fragment that survives to hit the ground is known as a **meteorite**.

Astronomers distinguish two types of meteors: the **sporadic,** whose orbital paths intersect that of Earth in random directions; and **shower meteors,** which are the remains of old comets that have left lots of small particles and dust in a common orbit. The material of sporadic meteors originates from the breakup of larger asteroids and old comets and the scattering of the debris away from the original orbits. When the orbit of shower meteors intersects that of Earth, numerous meteors may be viewed coming in from the same point, or **radiant,** in the sky. The association of meteors with comets is well known with the Leonids (observable around November 16 with a radiant in the constellation of Leo), representing the debris of Comet 1866I, and the Perseids (about August 11), which is the debris of Comet 1862III.

A typical meteor is only 0.25 grams and enters the atmosphere with a velocity of 30 km/s and a kinetic energy of approximately a 200,000 watt-second, allowing frictional heating to produce an incandescence equivalent to a 20,000 watt light bulb burning for 10 seconds. Daily, 10,000,000 meteors enter the atmosphere, equivalent to about 20 tons of material. The smaller and more fragile material that doesn't survive passage through the atmosphere is primarily from comets. Larger meteors, which are more solid, less fragile, and of asteroidal origin, also hit Earth about 25 times a year (the largest recovered meteorite is about 50 tons). Every 100 million years, an object 10 kilometers in diameter can be expected to strike Earth producing an impact that resembles the event that explains the demise of the dinosaurs at the end of the Cretaceous period. Evidence of some 200 large meteor craters remain preserved (but mostly hidden by erosion) on Earth's surface. One of the most recent and best known meteor crators that is preserved, the Barringer Meteor Crater in northern Arizona, is 25,000 years old, 4,200 feet in diameter, and has a depth of 600 feet. It represents an impact due to a 50,000-ton object.

Chemically, meteorites are classified into three types: **irons,** composed of 90 percent iron and 10 percent nickel), (representing about 5 percent of meteor falls), **stony-irons,** of mixed composition (1 percent of meteor falls), and **stones** (95 percent of meteor falls). The latter are composed of various types of silicates but are not quite chemically identical to Earth rocks. The majority of these stones are **chondrites,** containing **chondrules,** microscopic spherules of elements that appear to have condensed out of a gas. About 5 percent are **carbonaceous chondites,** high in carbon and volatile elements, and are believed to be the most primitive and unaltered materials found in the solar system. These meteorite classes provide evidence for the existence of chemically differentiated planetesimals (compare with the differentiation of the terrestrial planets), which have since broken up. Age-dating of meteorites yields the basic data for the age of the solar system, 4.6 billion years.

Asteroids

Asteroids, the largest non-planetary or non-lunar objects in the solar system, are those objects larger than 100 meters, or 1 kilometer, in diameter. The largest asteroid is Ceres, with a diameter of 1,000 km, followed by Pallas (600 km), Vesta (540 km), and Juno (250 km). The number of asteroids in the solar system increases rapidly the smaller they are, with ten asteroids larger than 160 km, 300 larger than 40 km, and some 100,000 asteroids larger than 1 kilometer.

The vast majority of asteroids (94 percent) are found between Mars and Jupiter in the **asteroid belt,** with orbital periods about the Sun of 3.3 to 6 years and orbital radii of 2.2 to 3.3 AU about the Sun. Within the asteroid belt, the asteroid distribution is not uniform. Few objects are found with orbital periods an integral fraction (½, ⅓, ⅔, and so on) of the orbital period of Jupiter. These gaps in the radial distributions of asteroids are called **Kirkwood's Gaps,** and are the result of accumulated gravitational perturbations by massive Jupiter, which altered the orbits to larger or smaller orbits. Cumulatively, the asteroids amount to a total mass of only 1/1,600 that of Earth and are apparently just debris left over from the formation of the solar system. Reflected sunlight from these objects shows that most of them represent three main types (compare with meteorites): those of predominantly metallic composition (highly reflective M-type asteroids, about 10 percent), those of stony composition with some metals (reddish S-type, 15 percent, and more common in the inner asteroid belt), and those of stony composition with high carbon content (dark C-type, 75 percent, more abundant in the outer asteroid belt). Asteroids with different proportions of silicates and metals come from the breakup of larger asteroidal bodies that once were (partially) molten, allowing chemical differentiation at time of formation.

Elsewhere in the solar system exist other groups of asteroids. The **Trojan asteroids** are locked into a stable gravitational configuration with Jupiter, orbiting the Sun at positions 60 degrees ahead or behind in its orbit. (These positions are known as the Lagrange L4 and L5 points, after the French mathematician who showed that given two bodies in orbit about each other, there are two other positions where a smaller third body may be gravitationally trapped). The **Apollo asteroids** (also called **Earth-crossing asteroids** or **near-Earth objects**) have orbits in the inner part of the solar system. These asteroids number a few dozen and are mostly about 1 kilometer in diameter. One of these small bodies likely will hit Earth every million years or so. In the outer solar system, we find the asteroid Chiron in the outer part of the solar system, whose 51-year orbit is probably not stable. Its diameter is between 160 and 640 kilometers, but its origin and composition are unknown. It may or may not be unique.

Comets

The structure of a typical **comet** includes gas and dust tails, a coma, and a nucleus (see Figure 4-6). The diffuse **gas** or **plasma tail** always points directly away from the Sun because of interaction with the solar wind. These tails are the largest structures in the solar system, up to 1 AU (150 million kilometers) in length. The tails are formed by sublimation of ice from the solid nucleus of the comet and look bluish due to re-emission of absorbed sunlight (fluorescence). Tail gases include compounds such as OH, CN, C_2, H, C_3, CO^+, NH_2, CH, and so on, for example, (ionized) fragments of ice molecules CO_2, H_2O, NH_3, and CH_4. A **dust tail**, appearing yellowish because of reflected sunlight, can sometimes be seen as a distinct feature pointing in a direction intermediate between the cometary path and the direction away from the Sun. The **coma** is the diffuse region around the nucleus of the comet, a region of relatively dense gas. Interior to the coma is the **nucleus**, a mass of mostly water ice with rocky particles (Whipple's dirty iceberg). Observation of the nucleus of Halley's Comet by spacecraft showed it to have an extremely dark surface, probably much like the dirty crust left on a snowpile melting in a parking lot. Typical cometary masses are about a billion tons with a size a few kilometers in diameter (Halley's Comet, for example, was measured to be an elongated object 15 kilometers long by 8 kilometers in diameter). Jets caused by gas boiling out of the nucleus sometimes can be observed, often forming an **anti-tail.** Jets can be a significant influence in changing a cometary orbit.

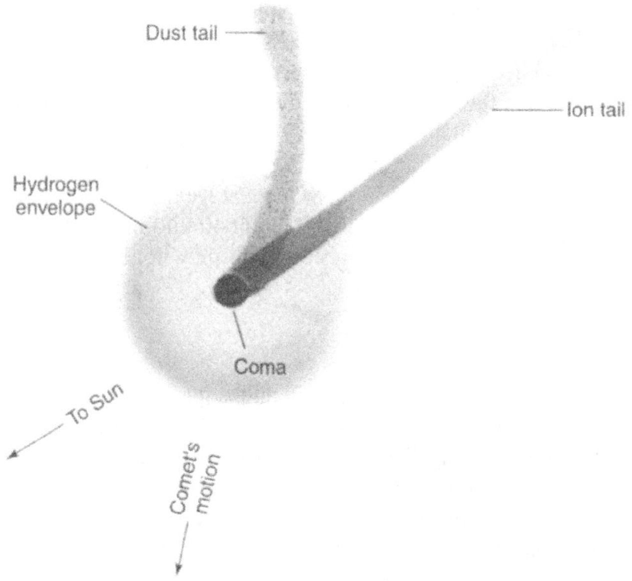

Figure 4-6

Schematic diagram of a comet.

Astronomers recognize two major groups of comets: **long period comets,** with orbital periods of a few hundred to a million years or more; and the **short period comets,** with periods of 3 to 200 years. The former comets have orbits that are extremely elongated and move into the inner solar system at all angles. The latter have smaller elliptical orbits with predominantly direct orbits in the plane of the ecliptic. In the inner solar system, the short period comets may have their orbits altered, specifically by the gravitation of Jupiter. There are about 45 bodies in Jupiter's family of comets with periods of five to ten years. Their orbits are not stable because of continued perturbations by Jupiter. In 1992, a dramatic perturbation between Comet Shoemaker-Levy and Jupiter occurred, with the comet breaking into some 20 fragments whose new orbit about Jupiter caused them to enter that planet's atmosphere some two years later.

Because comets are composed of ice that is slowly lost through solar heating, comet lifetimes are short compared to the age of the solar system. If a comet's perihelion is less than 1 AU, a typical lifetime will be about 100 orbital periods. The solid rocky material once held together by the ice spreads out along the cometary orbit. When Earth intersects this orbit, meteor showers occur. The finite lifetime of comets shows that a source of comets that continually supplies new ones must exist. One source is the **Oort Cloud,** a vast distribution of billions of comets occupying a region 100,000 AU in diameter. Occasionally, a comet is perturbed by a passing star, thus sending it into the inner part of the solar system as a long period comet. The total mass of the Oort Cloud is much less than that of the Sun. A second reservoir of comets, the source of the majority of short period comets, is a flattened disk in the plane of the solar system, but exterior to the orbit of Neptune. About two dozen objects with diameters of 50 to 500 kilometers have been detected in orbits out to 50 AU; but likely there are thousands more of these larger ones and millions of smaller ones in this **Kuiper Belt.**

Dust and gas

Dust and gas are the smallest constituents of the solar system. The presence of dust is revealed by its reflection of sunlight to produce the **zodiacal light,** a brightening of the sky in the direction of the plane of the ecliptic, which is best observed before sunrise or after sunset; and the **gegenschein** (or opposite light), again a brightening of the sky, but seen in the direction nearly opposite the position of the Sun. This brightening is caused by backscattered sunlight. Mapping of the sky by satellites using infra-red radiation has also detected thermal emission from bands of dust around the ecliptic, at the distance of the asteroid belt. The number of these dust belts agrees with the collision rate for major asteroids and the time for dust produced in such collisions to disperse.

Gas in the solar system is the result of the **solar wind,** a constant outflow of charged particles from the outer atmosphere of the Sun, which moves past Earth with a velocity of 400 km/s. This outflow is variable with a higher flux when the Sun is active. Exceptional flows of particles can cause disturbances in the magnetosphere of Earth, which can disturb long distance radio communication, affect satellites, and generate current anomalies in electrical power grids on the planet.

Other Planetary Systems

The first real observational evidence that other stars might produce planets came from infra-red observations coming from cool dust around stars. Study of several nearby stars, such as Vega (the brightest star in the summer sky) and β Pictoris, suggests the surrounding dust is in the form of disks, which could be similar to the protoplanetary nebula that surrounded the Sun prior to the nebula's concentration into a few planets. The star-forming regions in the direction of the Orion constellation show numerous young stars with surrounding material in the form of disks.

Present day technology does not allow astronomers to directly observe planets around other stars because the planets are lost in the glare of the central star. For instance, if Jupiter orbited the nearest star (other than the Sun), Jupiter's orbit would appear to an Earth observer only 3 arc seconds in radius. Jupiter itself would appear only 10^{-9} as bright as that star and would therefore be invisible to existing ground-based and satellite-based telescopes. To detect the planet's existence, astronomers could look for the 12-year oscillation in position (0.003") of the central star as the star and planet orbited around a mutual center of mass. Alternatively, observers could look for a change in the radial velocity of the central star as it moves about the center of mass. Such velocities would be small — of the order of 13 m/s — but are detectable. At present, this latter technique has identified some fourteen other stars with planetary-sized objects. (A planet is considered to have a mass less than about 13 times that of Jupiter. A mass greater than that would indicate an extremely faint, cool stellar object known as a **brown dwarf,** an object consisting of a body of gas that gives off a small amount of radiation but lacks sufficient mass to initiate the nuclear fusion that characterizes true stars.)

With such a limited number of planetary systems outside of our solar system to study, it is too early to draw major conclusions about the general nature of planetary systems and their process of origin. These discoveries of other planetary systems, however, do suggest that planets can form under unexpected circumstances. For example, two planetary systems have been found orbiting **pulsars,** the remnants of stars that exploded. Additionally, other planetary systems may display characteristics that differ substantially from the Sun's solar system, such as massive gas-giant planets extremely close to the central star and planets with large orbital eccentricities.

Several nearby stars, such as Vega (the brightest star in the summer sky) and β Pictoris, show evidence of surrounding dusty disks, which could be similar to the proto-planetary nebula that surrounded the Sun prior to the nebula's concentration into a few planets. The star-forming regions in the direction of the Orion constellation show numerous young stars with surrounding material in the form of disks.

CHAPTER 5
EARTH AND ITS MOON

Out of all the planets in the solar system, Earth is the only planet that scientists can study in detail. Atmospheric scientists can measure minute by minute atmospheric conditions (weather) from ground level to the "edge of space" by use of surface instruments and space vehicles. Geologists not only can detail surface features and how they change over time, but also can deduce Earth's structure to its very center. The division of Earth's interior into a core, mantle, and crust structure sets the context for how we study the other similar planets.

Properties of Earth and the Moon

Only a small number of physical factors actually distinguish the various objects in the solar system. There are numerical quantities like the total mass, a measure of the size (for spherical objects we use the radius), density, gravitational acceleration, and escape velocity. Other, more general terms can be used to indicate the present of an atmosphere, the condition of the surface, and the nature of the interior. Earth and its satellite, the Moon, compare as in Table 5-1.

Table 5-1: Properties of Earth and the Moon

Property	*Earth*	*Moon*	*Moon/Earth*
Mass	5.98×10^{24} kg	7.36×10^{22} kg	0.012 = 1/81
Radius	6370 km	1740 km	0.27
Mean density	5.5 g/cm^3	3.3 g/cm^3	0.61
Composition	More heavy elements	Less heavy elements	
Gravitational acceleration	9.8 m/s^2	1.6 m/s^2	0.165

(continued)

Table 5-1 ***(continued)***

Property	***Earth***	***Moon***	***Moon/Earth***
Escape velocity	11.9 km/s	2.4 km/s	0.21
Temperature range	220 K = –50°F (poles) 320 K = 120°F (deserts)	170 K = –150°F (night) 370 K = 212°F (day)	
Atmosphere	Yes	No	
Surface	Tectonic activity, Fast erosion	Geologically dead, Slow erosion	
Structure - core	Outer core liquid/ semi-liquid Inner core solid, 10-12 g/cm^3 Inner 0.53 of total radius	Solid, 7-8 g/cm^3 Inner 0.40 of total radius	
Structure-mantle	Mostly plastic, convecting 0.53 - 0.995 R	Mostly solid, fixed 0.40 - 0.96 R	
Structure-crust	Continental drift, erosion 0.005 R thick	Fixed 0.04 R thick	
Earthquakes	At crustal plate boundaries	At core/mantle transition	
Magnetic field	Yes	No	

Surface features

Topographically the Moon is very different from Earth. The Moon's surface is characterized by highlands and lowlands, mountains, and most notably, **craters** (bowl-shaped cavities of meteoric origin). These craters are often marked by secondary craters and by rays from **ejecta,** or ejected matter from the meteor's impact. The Moon's dark regions, called **maria,** are lava-filled basins up to 1,000 kilometers in diameter. Maria are sites of immense meteoric strikes early in lunar history that later were filled by molten lava seeping up from the interior. These maria are also the sites of gravity anomalies, or **mascons,** which are caused by the concentration of very dense material beneath the surface of the Moon. Mascons are found only on the near side of the Moon (the side of the Moon that faces Earth), suggesting that the influence of Earth's gravitation altered the trajectories of the impacting objects that produced these features.

Many of the lunar mountain ranges actually mark ancient crater rims. Unlike Earth, none of these features were formed by volcanism or by plate tectonic collisions. Rills and ridges that cross the lunar surface show evidence of surface contractions due to cooling of the rocky material of the lunar surface. The nature of the Moon's surface leads astronomers to the conclusion that it is basically original and was modified only by cratering and lava flows. By analyzing the Moon's physical features, therefore, we can deduce the early history of our solar system.

In contrast to the Moon, Earth's surface has an extremely varied topography. These differences can be attributed to two primary factors. First, as a larger object, Earth has cooled more slowly since it was formed. In fact, it is still cooling, with heat energy left over from the time of formation of Earth still slowly working its way outward. Energy always flows from hotter to cooler material; in Earth's interior the central heat in the core drives **convection currents** in the mantle that bring hot mantle material up toward the crust, and colder mantle and crustal rocks sink downward. At the Earth's surface this heat flow drives **plate tectonics (continental drift);** large segments of the earth's crust (plates) separated along deep cracks called **faults** are forced into motion. When the plates collide, these powerful internal tectonic forces squeeze and fold solid rock, creating massive changes in Earth's crust (see Figure 5-1). Mountain uplift and associated volcanic activity where plates collide are only two aspects of the continual recycling and rebuilding of the crust.

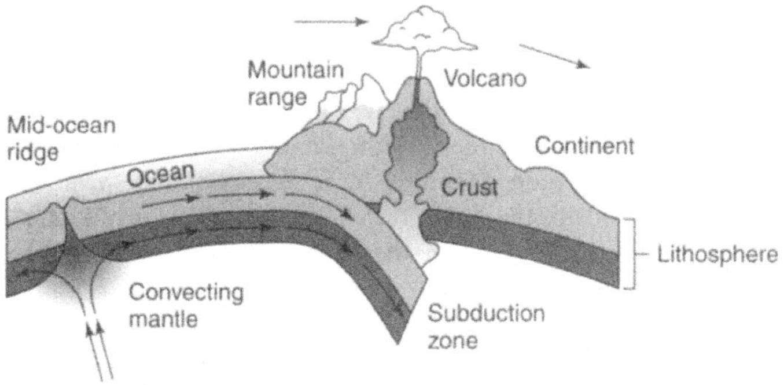

Figure 5-1

Earth's changing surface. Earth's surface is in a constant state of change due to factors such as convection currents, plate tectonics, and erosion.

The upwelling mantle material, driven by the flow of heat outward from the core of the planet, must spread out laterally beneath the crust, causing the continental plates to move apart. Because this movement occurs primarily in the denser surface rocks at the bottom of the oceans, it's termed **sea-floor spreading.** The weakened crustal structure allows molten material to rise, creating new surface rocks and **mid-oceanic ridges,** or mountain chains that can be traced for significant distances. The magnetic field patterns of oceanic sediments, symmetric on opposite sides of the mid-oceanic ridges, and the relative youth and thinness of mid-ocean sediments confirm continental drift. Researchers can also utilize radio astronomy techniques to directly measure motion showing, for example, that Europe and North America are drifting apart at a rate of several centimeters per year. The continents retain evidence of this drift, with shapes that resemble puzzle pieces that could be fitted together. The similarities between geological formations and fossil evidence show that indeed the present continents were once part of a single large land mass some millions of years ago.

Continental plates moving apart in one region means that elsewhere these plates must be colliding with other plates. Meanwhile, the denser ocean plates (heavier basalt) are moving under the lighter plates underlying the continental masses in **subduction zones.** These zones are marked by oceanic trenches, or mountain ranges caused by crumpling of continental materials to form mountain ranges, volcanism (for example, the Pacific ring of fire), and earthquake zones that obliquely dip below the continents.

Earth's surface is also constantly affected by the atmosphere (including wind and windblown sand and dust) and surface water (rain, rivers, oceans, and ice). Because of these factors, erosion of Earth's surface is an extremely rapid process. In contrast, the only erosive processes on the Moon are slow. There are the alternate heating and cooling of the surface during its month-long day; expansion and shrinkage only very slowly alter the surface. There are also impacts and slow modification of surface rocks from the solar wind.

Temperature and energy

The overall average temperature of Earth and the Moon (as well as any other planet) is due to a balance between the energy that they receive from the Sun and the energy that they radiate away. The first factor, energy received, depends on the planet's distance from the Sun and its **albedo** (A), the fraction of light reaching the planet that is reflected away and not absorbed. The albedo is 0.0 if all the light is absorbed and 1.0 for a if all the light is reflected. The Moon has an albedo of 0.06 because its dusty surface absorbs most of the light hitting the surface, but Earth has an albedo of 0.37 because clouds and the ocean regions are reflective. A planet's temperature may also be influenced by the greenhouse effect, or the warming of a planet and its lower atmosphere caused by trapped solar radiation.

The energy a planet receives per second per unit area (solar flux) is $L_{\odot}/4\pi R^2$, where $L_{\odot}$ is the solar luminosity and R is the distance from the Sun (residual heat coming from the planet's interior, energy produced from radioactivity, and humanity's combustion of fossil fuels have no significant effect on Earth's surface temperature). The total energy a planet absorbs per second is the fraction that is not reflected and also depends on the cross-sectional area of the planet, or $L_{\odot}/4\pi R^2 \times (1 - A)$. At the same time, the Stefan-Boltzman law ΣT^4 expresses the thermal energy emitted per second by each square meter of surface area (see Chapter 2). The total energy radiated per second is the Stefan-Boltzman Law times surface area, or $\Sigma T^4 \times 4\pi R(\text{planet})^2$. In equilibrium, there is a balance between the two, which yields the following: $L_{\odot}/4\pi R^2 = 4\Sigma T^4$. For Earth, this yields an expected temperature of T = 250 K = –9°F (a number lower than Earth's actual temperature because of the greenhouse effect).

On a microscopic level, energy absorption and energy emission is more complicated. Any small volume in the atmosphere is affected not only by the local absorption of solar energy, but also by the absorption of radiation from all other surrounding regions, energy brought in by convection (air currents), and energy gained by conduction (at the surface, if the ground is hotter). The loss of energy is due not only to the thermal black-body emission (see Chapter 2), but also by atomic and molecular radiation, energy taken away by convection, and energy removed by conduction (at the surface, if the air temperature is higher than the ground temperature). All these factors are responsible for the temperature structure of the atmosphere.

Earth's Atmosphere

Earth's atmosphere is divided into four regions. The lowest of these regions is the **troposphere.** Here exists turbulence and atmospheric mixing due to convection currents (weather, for example). The atmospheric density shows a rapid decrease with height. The dominant source of energy in this lowest region is conductive heating near the ground, and the dominant energy losses are radiation and convective mixing. The net result is a rapid temperature decrease with altitude. At a greater height is the **stratosphere,** where there exist strong horizontal wind currents (the **jet stream**) that affect ground level weather. Going higher, thc solar ultraviolet radiation that oxygen and the ozone absorb is the dominant source of energy gain, and radiation and convective mixing are the dominant means of energy loss. The balance between energy gain and loss results in a local temperature maximum at a height of 50 kilometers at the base of a region termed the **mesosphere.** Even higher is the **thermosphere,** where absorption of very short wave solar radiation is balanced by thermal radiation. Because there is very little material at this height to absorb this solar energy, the result is a high temperature. The lowest layer of the thermosphere (also called the **ionosphere,** 80 kilometers to 400 kilometers high) reflects radio waves because of ionization, and it is here that meteors burn up. Farther out, the extremely tenuous outermost part of the thermosphere (or **exosphere**) fades away until it is indistinguishable from the interplanetary material. (See Figure 5-2.)

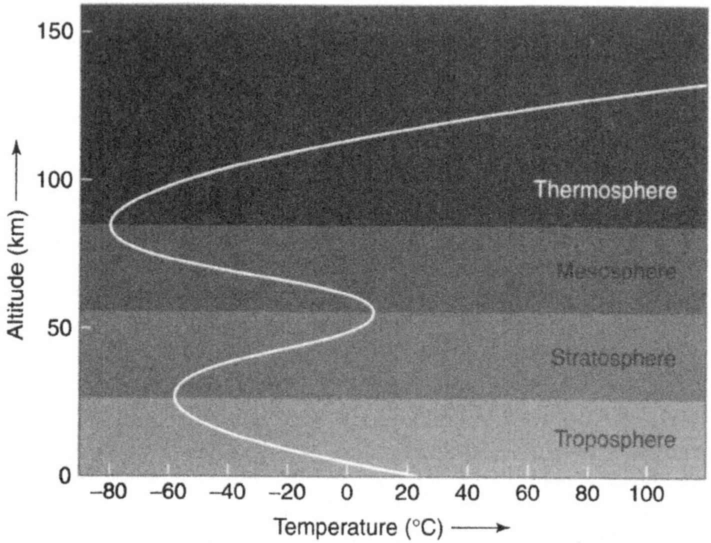

Figure 5-2

Temperature structure of Earth's atmosphere.

Global atmospheric conditions, such as weather, are also affected by the **Coriolis Effect.** Earth is a solid globe, and hence every latitude moves once around the rotational axis in the same period of time. The distance of travel around the rotational axis, however, depends on the latitude, with equatorial distance being the largest. The eastward velocity of Earth's surface, therefore, is greatest at the equator and least at the poles. Air masses moving north or south, however, share the eastward motion of that part of Earth's surface from where the air began moving and hence drift eastward or westward relative to Earth's surface. Air moving from all directions into a region of low pressure shows, relative to the surface, a counterclockwise motion in the Northern Hemisphere (hurricanes are the extreme examples), but clockwise in the Southern Hemisphere. Air moving in all directions outward from high-pressure regions circulates clockwise in the Northern Hemisphere.

The atmosphere further produces a thermal moderation of temperature over the whole Earth (resulting in less extreme temperatures both geographically and seasonally), shields the surface from life-destroying ultraviolet, and is the source of necessary gases for life.

The magnetosphere

The **magnetosphere** is that region surrounding Earth that is influenced by the planet's magnetic field (see Figure 5-3). Within it are two doughnut-shaped regions, the **Van Allen belts,** high above the equator, in which are trapped charged particles from the solar wind. These particles drift north and south along the magnetic field lines, enter the atmosphere near the magnetic poles, and produce the **aurora,** or the luminous irregular or streamer-like phenomena visible at night in a zone surrounding the magnetic poles. The solar wind streaming by the magnetosphere distorts it into a long tail pointing opposite the Sun. Other planets with magnetic fields also have similar magnetospheres.

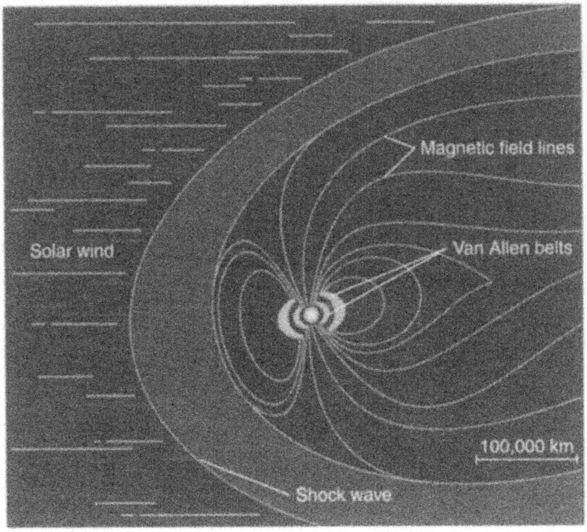

Figure 5-3

Earth's magnetosphere.

The greenhouse effect

Earth is warmer than expected from the global equilibrium between thermal radiation loss and absorption of sunlight because its atmosphere traps part of the thermal radiation that is trying to escape. This re-absorption of radiated thermal energy, or **greenhouse effect,** forces an increase in temperature in order to achieve a balance between solar heating and radiative energy loss (see Figure 5-4). Atmospheric gases that contribute to energy absorption are carbon dioxide (CO_2), water (H_2O), methane (CH_4), and chlorofluorohydrocarbons (freons). Consequently, Earth's surface temperature is closer to 300 K (80°F) than the expected 250 K. Venus, with its extremely dense CO_2 atmosphere, is the planet most dramatically affected by a greenhouse effect.

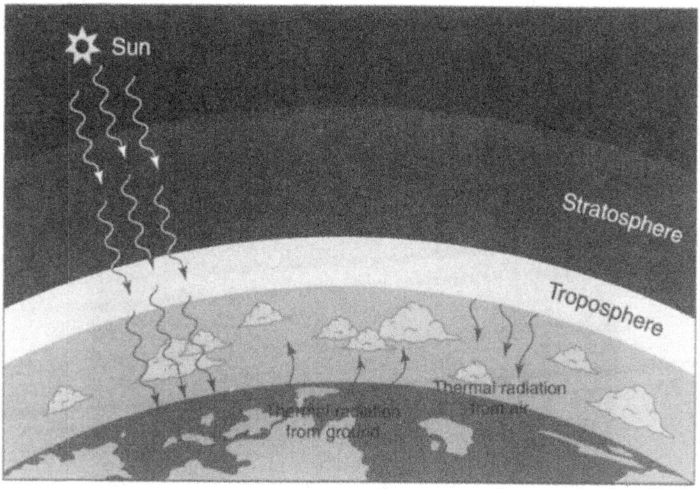

Figure 5-4

The greenhouse effect.

Earth's Chemical Composition

The composition of the atmosphere of Earth is 21 percent molecular oxygen, 78 percent molecular nitrogen, and 1 percent argon. Trace amounts of carbon dioxide, water vapor, and other gases are also present.

Direct chemical analysis of surface rocks shows their composition to be primarily oxygen, silicon, aluminum, and iron, in that order of abundance. Such rocks have an average density of about 2.7 g/cm^3, whereas the overall mean density of Earth is 5.5 g/cm^3, an observation that has two important consequences. First, the interior of the planet must have much denser materials than at the surface. The overlying weight of rock will compress interior rocks to some extent, but the necessary density requires the presence of intrinsically dense elements that must also be those that are cosmologically relatively abundant; that is, iron and nickel. Second, the differentiation of Earth's chemistry into lighter outer materials and heavier inner materials suggests that early in its history, the planet must have been reasonably molten in order to allow the heavier materials to sink into the interior.

Interior Structure — Core, Mantle, and Crust

The interior of Earth is not subject to direct investigation, but its properties must be indirectly deduced from the study of earthquake waves that propagate through the interior rocks. From an earthquake near the surface, both pressure (compression) waves and transverse (side-to-side) waves move outwards in all directions. Wave energy moving into the interior, however, has its path slowly changed by refraction as the wave moves through regions of slowly changing properties. These waves reach the surface after a time that depends on the length of the path and the velocity of propagation at each point along that path. Careful analysis at seismographic stations of the time of arrival of earthquake waves over the surface of Earth yields information about the densities, temperatures, and pressures of Earth's interior. A thin crust (at its thickest only 30 kilometers deep), which contains the continental masses and the ocean floors, overlies a denser outer **mantle.** The uppermost layer of the mantle acts as solid material, a **lithosphere** no more than about 80 kilometers deep. Most of the mantle slowly flows under pressure and acts as a plastic, or malleable, **asthenosphere.**

In an annulus about the surface of Earth, opposite an earthquake, exists the **shadow zone,** in which you cannot observe pressure waves. The path of pressure waves is significantly affected by a sharp refraction that astronomers interpret as the point of transition between the mantle and an interior **core** that is substantially different from the outer part of the planet. The shadow zone for transverse waves, however, covers the whole of Earth opposite the earthquake source. No transverse wave energy apparently passes through the core, indicating that its physical state, in the outer regions at least, must be liquid. The innermost core, however, though at higher temperatures, is likely solid because of an even higher pressure there. As the center of Earth continues to slowly cool over time, this inner core must be slowly growing in size at the expense of the liquid outer core. Evidence also shows that this inner core is rotating faster than the rest of the planet, completing one full turn in two-thirds of a second less time than at the surface. Applying other physical principles together with laboratory study of the nature of different materials under high temperature and pressure suggests the characterization of Earth's interior as shown in Table 5-2. (See Figure 5-5 for a diagram of Earth's interior.)

Table 5-2: Interior Properties of Earth

Property	*Crust*	*Mantle*	*Core*
Fraction of Earth	<1% of mass	~70%	~30%
State	"Broken rock"	Plastic	(Semi-)liquid
Depth (kilometers)	0-30	30-3030	3030-6370

(continued)

Table 5-2 *(continued)*

Property	*Crust*	*Mantle*	*Core*
Density (grams/ cubic centimeter)	2.7	3.5-5.5	10-12
Representative chemical composition	SiO_2	$(Fe,Mg)SiO_4$	Fe, Ni
Temperature (Kelvin)	300-500	500-3,000	3,000-5,300
Pressure (atmospheres)	1-1,000	10^3-10^6	10^6-10^7

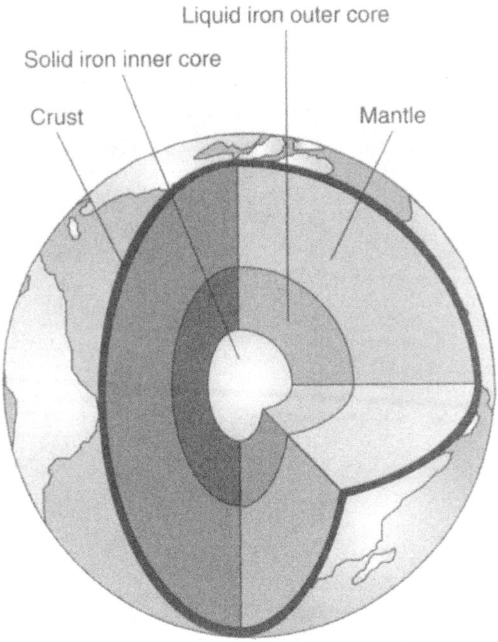

Figure 5-5

The interior of Earth.

Seismographic study of moonquakes has shown that the lunar structure is the same as the crust-mantle-core structure of Earth, with the significant differences being that the lunar mantle is primarily solid (the lunar lithosphere is about 800 kilometers deep and overlies only a shallow plastic asthenosphere), and the small iron core is frozen solid (see figure 5-6). As the Moon's mantle and core continue to slowly cool, their materials shrink at different rates, producing stress at the core-mantle interface; moonquakes thus occur in a deep spherical shell marking this interface. Because the Moon's outer mantle is frozen, unlike that of Earth, there is no interior convection, no surface plate tectonics, and no crustal quakes, other than an occasional tremor produced by the impact of a small meteor. In terms of interior structure, Earth and the Moon may be contrasted according to the information in Table 5-3.

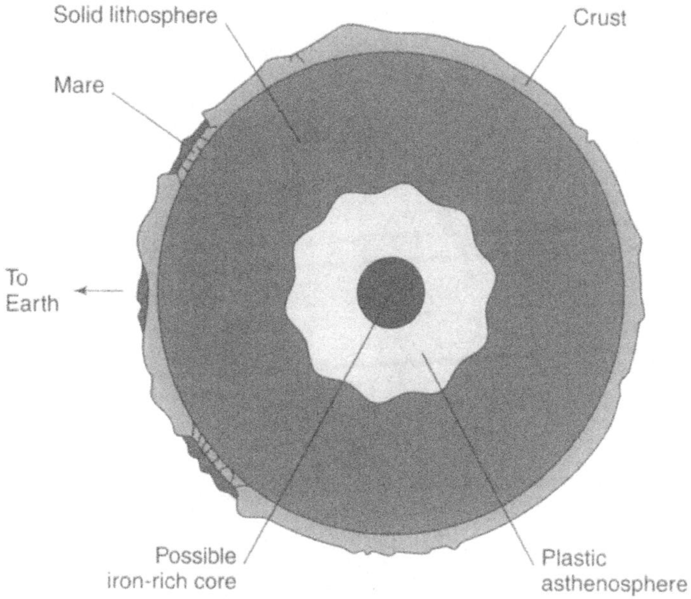

Figure 5-6

The interior of the Moon.

Table 5-3: Comparison of the Interior of Earth and the Moon

Property	*Earth*	*Moon*
Thickness of crust	30 km (0.5% of radius)	65 km (4% of radius)
Depth of mantle	30-3,000 km (48%)	65-1,200 km (67%)
Depth of core	3,000-6,370 km (inner 52%)	1,200-1,700 km (inner 29%)

The Age of Earth

The age of Earth (and by inference, the age of most other objects in the solar system) is also not directly known. But related evidence can be studied, in this case, by the technique of **radioactive dating.** Various elements (the parent element) are unstable and decay to produce another (the daughter) element. The time in which one-half of a parent sample decays into its daughter product is known as the **half-life** ($t_{1/2}$): it takes 4.5 billions years, for example, for one-half of a sample of uranium-238 (the form of uranium with 238 nuclear particles) to become lead-206. Alternatively, uranius-235 decays much quicker, with one-half of a sample becoming lead-207 in 710 million years.

After one half-life, the parent/daughter ratio is **one-half;** after two half-lives, the ratio is $(\frac{1}{2})^2 = \frac{1}{4}$, three half-lives, $(\frac{1}{2})^3 = \frac{1}{8}$, and so forth. Chemical analysis of a rock sample thus yields the present abundance ratios and an age for the formation of the rock. The oldest Earth rocks (which are rare due to the recycling of surface materials by plate tectonics) have an age of 3.8×10^9 years, which is a lower limit to the age of the planet and the solar system. A more correct estimate of the age of the solar system is based on materials that have been unaltered since their original formation. Applying radioactive dating to a specific class of meteorites believed to be unaltered since their formation yields consistent dates for their origin of $4.6 \pm 0.1 \times 10^9$ years. This solution is adopted as the age of Earth and the solar system.

Origin of the Earth-Moon System

The origin of the Earth-Moon system is very much related to the origin of the solar system as a whole. The ancient lunar surface has preserved a record of events over the last four billion years. Astronomers obtain relative crater ages from superimposition. For example, younger craters are found on top of older craters. Ejecta rays from younger craters also fall over older craters. Craters on lava flows (maria) similarly are younger than the lava. The purpose of the Apollo lunar missions was to obtain rock samples from different regions so that the relative age history of the lunar system could be translated into one with absolute ages. The planet Mercury, which is also heavily cratered with an apparently similar cratering history as the Moon, supplies additional evidence to theorize the Moon's history and origin. This, and other evidence, points to a process by which smaller objects (**planetesimals,** or little planets) merged to form the surviving planetary objects of today's solar system.

Earth and the Moon are so similar they can be thought of as forming a **binary planetary system.** Study of their chemical makeup provides important information on how these two objects became permanently associated with each other. The Moon is relatively deficient in heavier elements (mean density 3.3 g/cm^3 compared to 5.5 g/cm^3 for Earth). More specific chemical analysis of Moon rocks shows that the chemistry of the two objects is otherwise very similar, but not identical. Traditionally, three theories explain the association of the two objects. The theory of **coeval formation** argues that the Moon and Earth coalesced together out of the same materials. The idea that their chemistry is not identical poses a severe problem for this theory. **Fission theory** suggests that a single, initially rapidly rotating object broke apart. But this theory would require nearly identical chemical composition for the surviving objects. Dynamical problems also hinder this idea. The **capture hypothesis** theorizes that the Moon formed elsewhere in the solar system and only later became bound to Earth. This model allows for differences in the chemical composition of the two objects; but the problem is that their chemistry is too similar. Also, dynamical problems exist involving a loss of orbital energy necessary to end up with the two objects orbiting each other.

The ability of modern high speed computers to numerically model planetary-sized objects has led to a final theory that is likely correct—a grazing impact or **collisional hypothesis**. This theory provides that a Mars-sized object (a proto-moon about half the size of Earth) hit the proto-earth nearly tangentially. The proto-earth survived, but with significant crust/mantle material lost to a debris cloud surrounding the planet. The impactor was mostly disrupted into the debris cloud; its iron core survived more or less intact but was assimilated by Earth. Much of this debris (impactor mantle plus proto-earth mantle) subsequently coalesced to form the present Moon. Debris also fell to Earth to become part of its mantle and crust, thus producing lunar/terrestrial chemistry that is very similar, but not identical. Detailed computer calculations have shown that this scenario is dynamically and energetically possible.

Tidal Forces

The most direct consequence of the gravitational interaction between two objects is their mutual orbit about each other. However, the gravitational force also produces another effect because its magnitude is inversely proportional to the square of the distance and its directionality is between the centers of the masses. As a result, the gravitational force by one object on the *mass elements* of another depends upon where those elements are in the attracted object. Different elements of equal mass are pulled with different strengths and in different directions (see Figure 5-7a). What is really important, however, is the magnitude and direction of the gravitational force on each element *relative to the center of the attracted object*. These relative forces, or **tidal forces,** constitute a stretching in the direction of and a compression at right angles to the attracting object (see Figure 5-7b). The Moon, which keeps the same face to Earth during its monthly orbit, is stretched by about 20 kilometers along an axis pointing to Earth. The lunar effect on Earth is smaller because Earth is the more massive object. When the Moon is overhead or directly underfoot, the land surface is pulled about 1 foot higher by the Moon's tidal force. (Note that this means there are two high tides each day.) The oceans are more mobile, and hence ocean tides are typically about 6 feet high. Earth is subject not only to the lunar tidal influence, but also to a solar tide of approximately equal magnitude.

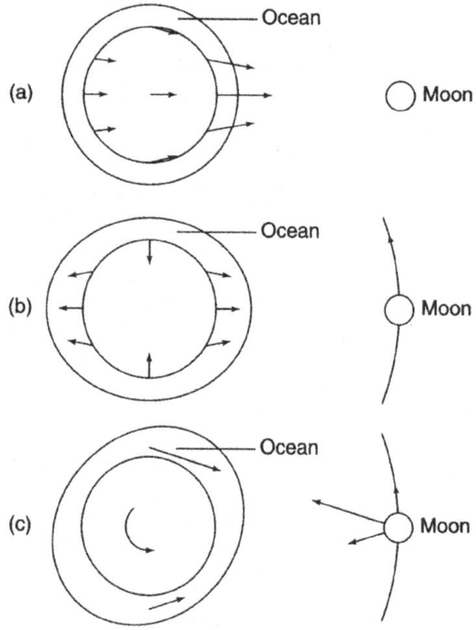

Figure 5-7

The effects of the Moon's tidal force on Earth.

Evolution of the Earth-Moon System

The tidal stretching of Earth is not an instantaneous effect. The mechanical strength of Earth's rocks produces a time delay in the tidal rise and fall of the solid surface. Similarly, it takes time for water to flow; hence, the ocean **tidal bulge** is not perfectly aligned to the direction of the Moon or Sun (see Figure 5-7c). The existence of the tidal bulge in turn results in additional gravitational forces that act opposite to the rotation of Earth, and in the direction of the motion of the Moon in its orbit. Earth's rotation, therefore, is slowing and the orbital distance to the Moon is slowly increasing with a proportional increase in its orbital period (the Earth-Sun effect is negligible, and hence the length of the year is basically constant). Both effects are measurable. Growth patterns in 400 million-year-old fossils show daily, monthly, and annual cycles suggesting an Earth day of 22 hours at that time and a lunar synodic month of 28 present days (in comparison to the current value of 29.5 days). The study of the occurrence of historical eclipses also shows the slowing of Earth's rotation. This slowing is also responsible for the annual or semi-annual addition of a second to our timekeeping in order to keep our clocks in synchronicity with Earth's rotation. Finally, direct measurement of the lunar distance over the last 25 years shows an annual increase in its orbital distance of about 2 centimeters per year.

It is equivalent to consider the tidal evolution of the Earth-Moon system in terms of energy. It takes energy to make Earth or the Moon stretch; thus rotational and orbital energy is expended or dissipated by tidal effects. This phenomenon is termed **tidal friction.** Bending a paper clip is analogous — mechanical energy must be used to bend the metal, which converts this energy into waste heat.

These tidal effects are mutual. The Moon acts on Earth, and Earth acts on the Moon. The Moon is the smaller object, and the effect of tidal friction has been to change the lunar rotation until its rotational period is equal to its orbital period about Earth — the Moon keeps the same face toward Earth. Ultimately, the Moon's action on Earth will produce a similar consequence. When Earth and the Moon achieve full synchronicity, with each rotation equal to their mutual orbital period, it is estimated that these periods will equal 55 present days and the Earth-Moon separation will be 613,000 kilometers. The effects of tidal evolution also are seen elsewhere in the solar system (see Mercury and Pluto in Chapter 6, for example).

Every moon in the solar system revolves with a period equal to its orbital period, thus keeping the same face to its primary planet. Both Pluto and its moon, Charon, have achieved synchronicity, each showing the same face to the other. And the planet Mercury, closest to the Sun, has a rotational period that matches its orbital motion about the Sun at perihelion, where the tidal forces are strongest.

CHAPTER 6
OTHER PLANETS OF THE SOLAR SYSTEM

The characteristics of the planets in the solar system indicate that really only two fundamental types of planets exist: those that are similar to Earth (the **terrestrial planets**) and those that are similar to Jupiter (**gas-giant planets**). The terrestrial planets include Mercury, Venus, Earth, and Mars. The gas-giant planets include Jupiter, Saturn, Neptune, and Uranus. The size and other properties of Pluto, however, make it unlike either of the terrestrial or gas-giant planets. Despite their similarities, however, planets are also like people—they have their own individual characteristics that distinguish each from all the other planets.

Mercury

Mercury is the hot, nearly airless planet that orbits closest to the Sun. It is about 1.5 times the size of the Moon and about 4.5 times the Moon's mass. The smallest of the terrestrial planets, it has an old (about four billion years), desolate, heavily cratered surface very much like the Moon, though its craters are flatter and thinner rimmed because of the planet's higher surface gravity. Also present on Mercury's surface are large multi-ringed lava-filled craters, called **basins**, and long, high cliffs, or **scarps,** produced either during the early cooling and contraction of the planet or by the solar tidal effect. The impact that produced the largest basin, **Caloris,** also sent shock waves through the interior of the planet; these converged on the planet's opposite side to produce an exceedingly rugged surface. Evidence also shows extensive volcanic events, in the form of younger **intercrater plains** and **smooth plains,** with fewer and smaller craters than adjacent areas.

The planet has a surprisingly high mean density, 5.4 g/cm^3, suggesting that its iron-nickel core is disproportionately large, possibly due to partial loss of its mantle in a collision with another large object (see Figure 6-1). This iron core is responsible for its weak magnetic field (1 percent that of Earth), likely a residual field left over from an early time when more rapid rotation generated electrical currents in the core, which in turn generated the magnetic field. Its proximity to the sun produces an extreme temperature range between 700 K (760°F) during its daytime and about 100 K (–280°F) during its night. Ice appears to exist in the permanently shadowed interiors of deep craters near the poles. Little atmosphere exists—about one-trillionth that of Earth, and is composed of solar wind gases temporarily held by the planet's gravity and some sodium and potassium likely released from surface rocks by the impact of the solar wind particles.

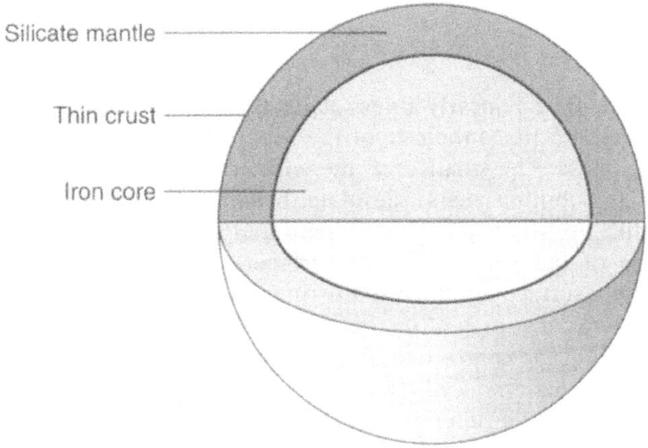

Figure 6-1

The interior of Mercury.

Of the inner eight planets, Mercury has the largest orbital eccentricity (e = 0.21). Its large variation of distance from the Sun has produced, via tidal friction (see Chapter 5), an unequal **resonance,** or relation between its rotation and its orbit. With a 59-day rotation and 88-day orbit, Mercury's resonance has a 3:2 ratio, or three rotations in every two orbits about the Sun. The combination of its rotation and revolution about the Sun produces a Mercury day equal to two orbital periods, or 176 Earth days. Careful consideration of the dynamics of the planet shows that this is the result of **tidal locking** at perihelion, when the planet is closest to the sun and the tidal forces are strongest.

Mercury has also been the object of study because its orbit cannot be adequately explained by only using Newton's Law of Gravity. According to Newton, the gravitational force between a spherical Sun and Mercury produces an elliptical orbit that perfectly repeats itself each time the planet moves around the Sun (Kepler's Laws). Because the other planets also gravitationally tug on Mercury and because the Sun is not perfectly spherical, each successive elliptical path is actually rotated slightly with respect to the previous orbit, like a spirograph pattern. The position of the closest approach of Mercury (its perihelion) thus slowly drifts around the Sun (an effect termed **precession of the perihelion**) at the rate of 5600"/century = 1.56°/century = 13.5"/orbit. When the effect of the non-spherical Sun and the gravitational perturbations by the other planets are subtracted, there is left over a precession (rotation) of 43"/century = 0.1"/orbit, a tiny change that can be accounted for only by Einstein's theory of general relativity, but not by Newton's theory of gravitation. Mercury's precession is thus one of the fundamental tests showing that the theory of general relativity is a far more accurate description of gravitational phenomena that is the Newtonian theory, although the difference in predictions between the two is negligible under most circumstances. See Table 6-1 for Mercury's physical and orbital data.

Table 6-1: Mercury

Physical Data	
Diameter (equatorial)	4879 km
Oblateness	0.0
Inclination of equator to orbit	0°.0

(continued)

Table 6-1 ***(continued)***

Physical Data	
Axial rotation period (sidereal)	58.646 days
Mean density	5.43 g/cm^3
Mass (Earth = 1)	.06
Volume (Earth = 1)	.06
Mean albedo (geometric)	0.11
Escape velocity	4.25 km/s
Orbital Data	
Mean distance from Sun (10^6 km)	57.909
Mean distance from Sun (AU)	0.387
Eccentricity of orbit	0.206
Inclination of orbit to ecliptic	7°.0
Orbital period (sidereal)	87.969 days

Venus

In size and mass, Venus is a near twin to Earth, but in other characteristics, it shows significant differences. Compared to the other terrestrial planets, Venus' atmosphere is extremely dense, with a surface pressure 90 times that of Earth, equivalent to 1,300 pounds per square inch. Chemically, it is 96 percent carbon dioxide (CO_2) and slightly less than 4 percent molecular nitrogen (N_2), with the remainders being argon (Ar) and water vapor (H_2O), and trace amounts of sulfuric acid (H_2SO_4), hydrochloric acid (HCl), and hydrofluoric acid (HF). The enshrouding clouds, which historically prevented Earth-based study of the planet, are most likely droplets of sulfuric acid (H_2SO_4), a product of the planet's active volcanism. The planet's temperature at the cloud level is 240 K (–35°F) as expected for its high

albedo (reflectivity) of 0.76; but at the surface, the temperatures are far higher—750 K (890°F) on the daytime side and 550 K (500°F) on the night side. The planet's dense carbon dioxide atmosphere is responsible for an extreme greenhouse effect. The large temperature difference between the planet's day and night sides produces extremely high winds in the upper atmosphere—300 km/hr, which can circle the planet in only four days. (In contrast, on Earth, the solar heat is fairly uniformly spread about planet.) (See Figure 6-2.)

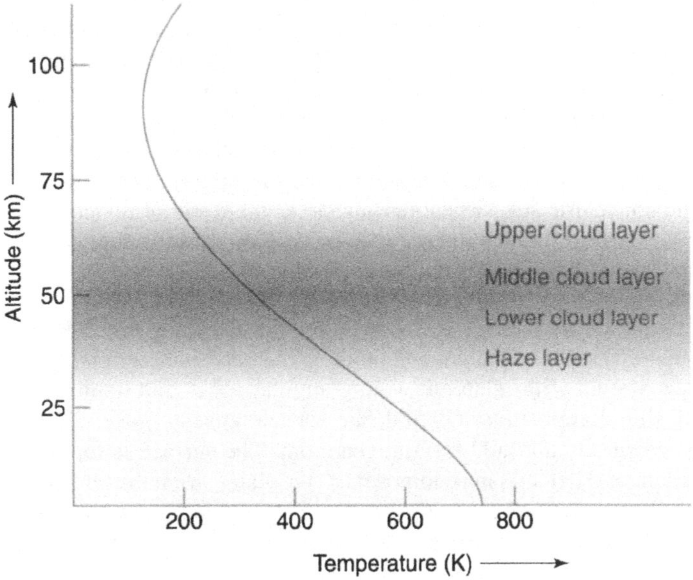

Figure 6-2

Temperature in Venus' atmosphere.

That Venus is so close to Earth in size and mass makes the difference in atmosphere even more striking. Why should they be so different in this respect? Actually the atmosphere of Venus contains the quantity of carbon dioxide (CO_2) expected from volcanic processing (**volcanic outgassing**) of the rocky material that makes up that body.

It is therefore Earth that has the odd atmosphere due to the existence of the oceans and also due to life on the planet. The oceans dissolve CO_2 and, augmented by processes of living organisms, precipitate this out of a solution in the form of solid limestone. If all CO_2 were released from the ocean and rocks, the terrestrial atmosphere would be like that of Venus.

The surface of Venus can only be mapped by radar techniques, with low resolution from Earth and with higher resolution (1 km) achieved by the Magellan satellite that was placed in orbit about the planet. This has revealed the existence of highland ("continental") and lowland (basin) regions; but the elevation contrast is relatively shallow compared to Earth and better described as a rolling terrain. Meteoric impact craters, volcanic caldera, and a number of other surface features quite unlike those known on Earth are found in Venus' topography. For example, coronae, circular bulges marked by radial and concentric stress fractures, appear as the result of the movement of subsurface molten materials. Other regions show patterns of parallel ridges suggesting some compression of these areas.

Active vulcanism was confirmed by the short-lived Russian Venera spacecraft that landed on the planet and photographed a surface strewn with relatively young angular rocks and boulders of a basaltic composition. Overall, the surface appears generally to be between 300 and 500 million years old. The surface is too hot for oceans to exist, and in its atmosphere the water would be quickly disassociated due to the subsequent loss of hydrogen. The isotopic abundance ratio of the remaining deuterium (heavy hydrogen) and hydrogen, however, shows that Venus originally had some water, but not enough to have formed extensive oceans. Without oceans, there was no dissolution of atmospheric carbon dioxide (CO_2) and no precipitation out in the form of solid limestone. As the surface temperatures rose, the extreme heating of the surface rocks released even more CO_2 and the greenhouse temperature increase grew even higher.

Also by comparison with Earth, Venus should still be actively cooling and the features of active plate tectonics phenomena should be found on its surface, but they are not. There are no extended mountain chains equivalent to Earth's mid-oceanic ridges, no large scale faults associated with sea floor spreading, no subduction troughs, and no marginal mountains caused by crustal compression at the collision of continental plates. The mantle of Venus may once have been convecting, but now it and the crust appear to be frozen solid. The high surface temperature has two consequences for the dynamics of the planet's mantle. First, the temperature difference between core and surface is smaller than on Earth, hence there is a smaller thermal driving force to produce convection. Second, the high outer temperature "cooked" the mantle, driving out water that otherwise would be chemically bonded to silicate materials. Without the water, the mantle is solid, not plastic (in the same way that silly putty placed in an oven will lose its elastic properties).

The rotation of the planet is unusual, being the slowest (244 days) of any planet and opposite in direction (that is, retrograde) to the rotations of most solar system objects. This odd rotation is likely the result of an early impact by a large planetesimal during the planet's formative period; or perhaps Venus was the final result of the merger of two large planetesimals. The length of Venus' solar day, consequently, is 117 Earth days, hence the strong differential heating between its day and night sides. See Table 6-2 for Venus's physical and orbital data.

Table 6-2: Venus

Physical Data	
Diameter (equatorial)	12.104 km
Oblateness	0.0
Inclination of equator to orbit	177°.36
Axial rotation period (sidereal)	243.02 days
Mean density	5.24 g/cm^3
Mass (Earth = 1)	0.82
Volume (Earth = 1)	0.86

(continued)

Table 6-2 ***(continued)***

Physical Data	
Mean albedo (geometric)	0.65
Escape velocity	10.36 km/s
Orbital Data	
Mean distance from Sun (10^6 km)	108.209
Mean distance from Sun (AU)	0.723
Eccentricity of orbit	0.007
Inclination of orbit to ecliptic	3°.4
Orbital period (sidereal)	224.701 days

Mars

An intermediate-sized terrestrial planet, Mars has a cratered, reddish surface that is now structurally inactive, as the last active vulcanism ended some 1.5 billion years ago. Although it has no global magnetic field, mapping its patterns of surface magnetism reveals parallel strips of land with opposite magnetic polarity, analogous to the magnetic patterns found on Earth on either side of the mid-oceanic ridges. These strips of land suggest that very early in its history, before it had cooled significantly, Mars had an inner molten core that was able to generate a magnetic field and an outer mantle that was convecting to produce plate tectonic phenomena at the surface. Today, the planet's surface preserves the greatest range of elevation of any of the terrestrial objects, from a 1,300 kilometer-wide crater in the northern hemisphere, whose floor is 10 kilometers below mean elevation, to the 25 kilometer-high Olympus Mons, the largest volcano in the solar system. Olympus Mons is an example of a **shield volcano,** formed when thin lava spreads out over a large area. The diameter of this extinct volcano is 600 kilometers, or three times the wide size of the largest

Earth volcano of this type (located in Hawaii). Volcanic resurfacing of the planet affected primarily the planet's northern hemisphere, which is covered with relatively smooth plains. The older southern hemisphere surface shows the accumulation of many more craters over time.

The cratered regions show the influence of weathering. The uncratered regions, large "deserts" covered with sand dunes, show seasonal changes in appearance as the annual cycle of winds moves the loose dust from one area to another. Canyon areas are apparently due to erosion by wind, old lava scouring of channels, and/or water, although no liquid apparently exists on the planet's surface today. Remote satellite mapping of the planet's surface is becoming more detailed, with increasing evidence that surface water flow has left characteristic erosion channels. A major equatorial canyon, Valles Marineris, is 4,000 kilometers long, with tributary canyons. It was likely caused by uplift (a bulge formed by internal movement of deep subsurface molten material) and splitting of the Tharsis region on the planet (not a rift valley as found on Earth). The existence of teardrop-shaped islands, ancient river channels, and flood plains all suggest flow of water on the surface, but any water now on the planet must be locked into the surface rocks in the form of a permanent tundra or permafrost.

The ice caps wax and wane with the seasons (globally, the temperature is cold, averaging 220 K = −60 °F, with ± 30 K seasonal variation). These ice caps are composed primarily of carbon dioxide and possibly some water, representing a substantial, frozen fraction of the planet's thin atmosphere. As a consequence, the atmospheric dynamics of Mars are far different from that of Earth. In the Martian northern hemisphere spring, the northern ice cap begins to shrink with the liberated carbon dioxide moving south only to be frozen into the southern hemisphere ice cap. At the onset of the southern hemisphere spring, the process is reversed. In spite of its thinness, the atmosphere can produce high velocity winds and planet-wide dust storms. Dust settling onto the ice caps is left behind when the ice sublimates each spring, forming more or less concentric terraced areas about the poles.

This Martian atmosphere has a surface pressure about 1 percent of that of Earth, far less than expected based on the planet's volcanic past. Because the planet's gravity is sufficient to have retained its atmosphere, another cause must exist to explain its lack of an atmosphere. It has been suggested that most of the original atmosphere may have been lost in a meteoric collision event long ago or possibly eroded away by interaction with the solar wind. Compositionally, what little air there is consists of 95 percent carbon dioxide, 3 percent nitrogen, and 1.5 percent argon, with seasonal trace amounts of water, oxygen, and helium. Its density falls off with altitude more slowly than on Earth, thus at a height of 50 kilometers, the Martian atmosphere is actually denser than that on Earth at the same height. The atmosphere thus can support high clouds, generally in the morning. (See Table 6-3 for Mars's physical and orbital data.)

Internally, the planet's structure follows the characteristic crustmantle-core structure. The planet's low mean density of 3.9 g/cm^3 suggests a relatively small iron core. No evidence exists for current tectonic activity.

Two small irregularly shaped moons, Phobos (roughly 28 km by 23 km by 20 km) and Deimos (approximately 16 km by 12 km by 10 km), orbit the planet, each keeping its long axis always pointing toward the planet for the same reason that the Moon keeps the same face toward Earth. Both are heavily cratered and likely are captured asteroids, not objects that formed in conjunction with the formation of the planet. Phobos (the one in the lowest orbit) is slowly spiraling in toward the planet because of atmospheric friction and is predicted to crash to the surface within the next 100 million years.

Table 6-3: Mars

Physical Data	
Diameter (equatorial)	6,794 km
Oblateness	0.0065
Inclination of equator to orbit	25°.19

Physical Data	
Axial rotation period (sidereal)	24.623 hours
Mean density	3.94 g/cm^3
Mass (Earth = 1)	0.11
Volume (Earth = 1)	0.15
Mean albedo (geometric)	0.15
Escape velocity	5.02 km/s
Orbital Data	
Mean distance from Sun (10^6 km)	227.941
Mean distance from Sun (AU)	1.524
Eccentricity of orbit	0.093
Inclination of orbit to ecliptic	1°.9
Orbital period (sidereal)	686.980 days

Jupiter

Jupiter is the largest planet in the solar system and contains two-thirds of the mass found outside the Sun in the solar system. It appears reddish, with well-defined **belts** (darker zones) and **bands** (brighter zones) encircling the planet. These features do not represent surface area, but are clouds. The bands are tops of an upwelling of light colored, warm gas from the planet's interior. In the upper atmosphere, these gases react with solar ultraviolet radiation, forming dark hydrocarbons or photochemical smog. At the same time, these gases cool, become more dense, and sink back into the interior. The planet's rapid rotation (its period is about 10 hours) not only produces a significant equatorial bulge, but also a strong Coriollis effect, giving gases moving north or south additional westward or eastward motion, thus distorting the convective regions into latitudinal zones that wrap around the planet. The Coriolis effect also produces the cyclonic storms in the planet's atmosphere, including the immense **Great Red Spot,** which has survived for at least 400 years. (The Spot is 35,000 km diameter, has a top that is 8 km higher than surrounding cloud layers, and rotates at 500 km/hr; it also drifts somewhat in position.)

Chemically, the outer atmosphere (see Figure 6-3) is composed of molecular hydrogen (60 percent by mass), helium (36 percent), neon (2 percent), water vapor (0.9 percent), and lesser amounts of ammonia, argon, and methane. The reddish color is due to acetylene formation and the release of phosphorous from phosphine. In the upper atmosphere, distinct cloud layers form from ammonia, water, and ammonium hydrosulfide.

The outer few hundred kilometers of the planet were explored by the Galileo probe in 1995 and its physical characteristics deduced from the impacts of the fragments of Comet Shoemaker-Levy in the previous year. The internal structure of the planet, however, cannot be directly observed and must be deduced from its mean density of 1.3 g/cm^3 and applicable physical laws. In the outer part, the planet is gaseous, and the balance between gas pressure and gravity is the major factor in setting its structure. At high enough pressure, molecular hydrogen is dissociated into individual atoms. But this same high pressure also means that the electrons are no longer tightly bound to individual hydrogen atoms, but can easily move from one atom to another allowing the generation of electrical currents. The ability to permit an electrical current is a property of metals, hence this state of hydrogen is termed **metallic atomic hydrogen**. The electrical currents in turn produce a strong magnetic field inclined 10 degrees to the planet's rotational axis. Exterior to the planet there are super versions of Earth's van Allen belts in which charged solar particles are trapped. As these move along the magnetic field lines and enter the upper atmosphere in the regions of the magnetic poles, strong aurorae are produced.

Much closer to the center of the planet, under higher temperatures and pressures, the possible existence of ices, as well as a solid core, must be considered. The interior of Jupiter is not as well known as astronomers would like, for they must make various assumptions in the effort to calculate interior conditions; slightly different assumptions lead to some variance in the computed interior conditions.

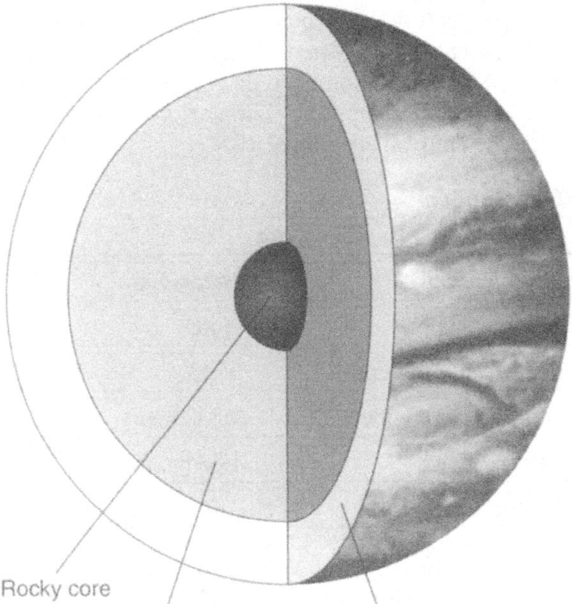

Figure 6-3

The interior of Jupiter.

The temperature of the upper atmosphere of Jupiter (125 K) is set by the balance between its absorption of solar energy and its thermal radiation. Jupiter also emits a large amount of radiation in the radio region, produced by lightning in the atmosphere and from

Jupiter's magnetic interaction with its moon Io (see the section on Moons in Chapter 4). Considering all forms of energy emission, Jupiter radiates about two times as much energy as absorbed from sunlight; thus it must have an internal energy source. Powerful convection keeps its outer composition well mixed, thus the only possible energy source is gravitational contraction. A slow shrinking of the planet, about three centimeters per century, is adequate to produce the excess energy. (Alternatively, it can be considered that the planet is still losing its primordial heat. As it cools, gas pressure slowly decreases relative to gravity and thus the planet continues to shrink slowly in size.) See Table 6-4 for Jupiter's physical and orbital data.

Table 6-4: Jupiter

Physical Data	
Diameter (equatorial)	142,984 km
Oblateness	0.065
Inclination of equator to orbit	3°.13
Axial rotation period (sidereal)	9.842 hours
Mean density	1.33 g/cm^3
Mass (Earth = 1)	317.8
Volume (Earth = 1)	1323
Mean albedo (geometric)	0.52
Escape velocity	59.6 km/s
Orbital Data	
Mean distance from Sun (10^6 km)	778.328
Mean distance from Sun (AU)	5.203
Eccentricity of orbit	0.048
Inclination of orbit to ecliptic	1°.3
Orbital period (sidereal)	11.862 years

Saturn

Saturn, the second largest planet, is basically a smaller version of Jupiter, an obvious difference being the spectacular system of thousands of concentric, thin rings that encircle the planet. Being farther from the Sun, its outer atmosphere is cooler (95 K), and hence its cloud layers occur much deeper within its hazy atmosphere; its belts and bands thus appear washed out. Like Jupiter, its atmospheric composition is mostly hydrogen (80 percent) and helium (20 percent). Compared to Jupiter, helium appears depleted in Saturn, suggesting that convection is unable to keep the composition well mixed. The sinking of heavier helium into the planet's interior may be the source of the excess energy that the planet emits over and above the solar energy it absorbs. In its interior (see Figure 6-4), hydrogen occurs mostly in its molecular form with a smaller metallic hydrogen zone near the center (and hence a weaker magnetic field). A region of solid ice around the core is more likely than for Jupiter, but at best, there is only a tiny iron-silicate core for the overall density of the planet, which is only 0.7 g/cm^3. (See Table 6-5 for Saturn's physical and orbital data.)

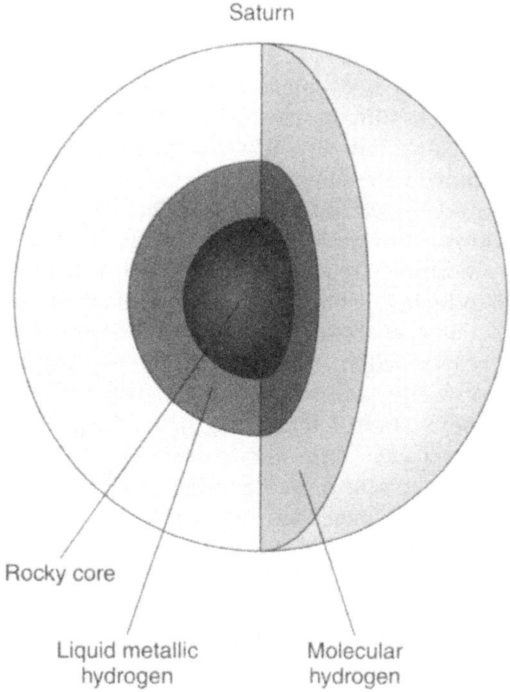

Figure 6-4

The interior of Saturn.

Table 6-5: Saturn

Physical Data	
Diameter (equatorial)	120,540 km
Oblateness	0.098
Inclination of equator to orbit	26°.73
Axial rotation period (sidereal)	10.233 hours
Mean density	0.70 g/cm^3

Physical Data	
Mass (Earth = 1)	95.16
Volume (Earth = 1)	752
Mean albedo (geometric)	0.47
Escape velocity	35.6 km/s
Orbital Data	
Mean distance from Sun (10^6 km)	1426.990
Mean distance from Sun (AU)	9.539
Eccentricity of orbit	0.056
Inclination of orbit to ecliptic	2°.5
Orbital period (sidereal)	29.457 years

Uranus

Although another gas-giant planet whose deep atmosphere is composed primarily of hydrogen (74 percent) and helium (26 percent), Uranus appears as a nearly featureless greenish ball because of methane in the outer atmosphere and an apparent lack of strong convection. At a temperature of 65 K, ammonia and water, which are responsible for the clouds of Jupiter and Saturn, have frozen out and "snowed" into the planet's interior. The gaseous methane that remains behind in the upper atmosphere absorbs red light; hence sunlight, reflected back into space, gives the planet its greenish color.

In size (four times that of Earth) and mass (14 times that of Earth), Uranus is almost a twin of the planet Neptune; though slightly less massive than Neptune, it is slightly larger in radius because of its weaker gravity. Compared to Jupiter, Uranus has a relatively larger iron/rocky core (see Figure 6-5). Its major distinguishing feature is a rotational axis nearly in the plane of the solar system, a factor that certainly must have been a product of its formation (it was affected either by a major collision or resulted from the merger of two nearly equal bodies). In contrast to the three other gas-giant planets, Uranus is in balance between the energy it emits and the energy it absorbs from sunlight. Without internal heat moving slowly outwards by convection, no equatorial cloud bands can be produced, even though the planet's rapid rotation is accompanied by a strong Coriolis effect. (See Table 6-6 for Uranus's physical and orbital data.)

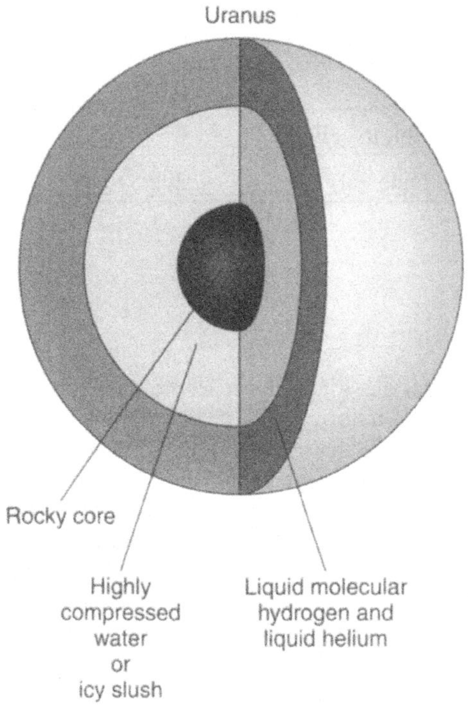

Figure 6-5

The interior of Uranus.

Table 6-6: Uranus

Physical Data	
Diameter (equatorial)	51,118 km
Oblateness	0.023
Inclination of equator to orbit	97°.86
Axial rotation period (sidereal)	17.24 hours
Mean density	1.3 g/cm^3
Mass (Earth = 1)	14.5
Volume (Earth = 1)	64
Mean albedo (geometric)	0.51
Escape velocity	21.1 km/s
Orbital Data	
Mean distance from Sun (10^6 km)	2,869.549
Mean distance from Sun (AU)	19.182
Eccentricity of orbit	0.047
Inclination of orbit to ecliptic	0°.8
Orbital period (sidereal)	84.01 years

Neptune

With an atmospheric temperature of 55 K, this near twin to Uranus appears deeply blue because of the even greater content of methane in its outer atmosphere. Very faint belts and zones are present (there is convection in the outer part of the planet), as well as cyclonic storms, including the **Great Dark Spot,** a storm the size of Earth. High, whitish atmospheric clouds of methane also occur.

Neptune is slightly more massive than Uranus; hence its somewhat greater gravity has caused greater compression of its atmosphere, thus it appears smaller. Like Jupiter and Saturn, it radiates more energy than it receives from the Sun. This excess energy loss must come from its primordial heat and thus it must be shrinking slightly. (See Table 6-7 for Neptune's physical and orbital data.)

Table 6-7: Neptune

Physical Data	
Diameter (equatorial)	49,528 km
Oblateness	0.017
Inclination of equator to orbit	28°.31
Axial rotation period (sidereal)	16.11 hours
Mean density	1.76 g/cm^3
Mass (Earth = 1)	17.20
Volume (Earth = 1)	54
Mean albedo (geometrik)	.41
Escape velocity	24.6 km/s
Orbital Data	
Mean distance from Sun (10^6 km)	4,496.637
Mean distance from Sun (AU)	30.058
Eccentricity of orbit	0.009
Inclination of orbit to ecliptic	1°.8
Orbital period (sidereal)	164.793 years

Pluto

Pluto is the outermost of the known planets, in an atypical orbit that is both eccentric (e = 0.25) and significantly inclined (17 degrees) to the plane defined by the orbits of the inner planets. At perihelion, it is actually closer to the Sun than the planet Neptune. It is the only planet that cannot be placed in either the terrestrial or gas-giant classification. Overall, in size (with a diameter of 2,290 km, Pluto is the smallest of the known nine planets), mass (0.2 percent that of Earth, or about one-sixth that of the Moon), and density (about 2 g/cm^3, indicating a composition of roughly half ice and half silicate materials), it is more like a moon of one of the outer planets or the largest and closest of a belt of comet-like objects orbiting beyond Neptune.

Pluto has its own relatively large and close moon, Charon. Pluto and its moon therefore form another binary planetary system with a period of 6.4 days and in an orbit that is tilted 106° to their orbit about the Sun. Both rotate synchronously with their mutual orbit; hence they keep the same faces toward each other. Spectroscopy shows that methane is present as surface ice (the planet's temperature is 45 K) and as a thin atmosphere (with nitrogen N_2, the surface pressure is only 10^{-6} that of Earth's atmosphere) that possibly surrounds both the planet and its moon. The planet's surface reflectivity is not uniform, but this changes over time. As the planet moves outward from the Sun, its temperature falls and the atmosphere freezes out, falling as snow on the surface of the planet and increasing its reflectivity. When it moves inward, closer to the Sun, solar heating causes the surface ice to return to its gaseous state in the atmosphere.

The unusual physical properties of Pluto are matched by its unique orbital relationship to Neptune. Pluto orbits the Sun twice during three Neptune orbits. This results in an unusual circumstance: Whenever Pluto is closest to Neptune's orbit, Neptune is always on the other side of the Sun. In fact, if Pluto were not in this orbital resonance with Neptune, it is likely that long ago it might have been gravitationally expelled from the solar system.

All these properties suggest that Pluto is not a true planet in the sense that the other planets are — namely, that they are objects that are the result of gravitational assemblage of prior small objects, or planetesimals. Rather, Pluto (and its moon Charon) are left over pieces of outer solar system planetesimals, some of which were accreted into the major planets, others of which became moons, and others that were lost to the solar system in gravitational interactions with other planets. (See Table 6-8 for Pluto's physical and orbital data.)

Table 6-8: Pluto

Physical Data	
Diameter	2,302 km
Oblateness	0
Inclination of equatior to orbit	122°.5
Axial rotation period (sidereal)	6.387 days
Mean density	1.1 g/cm^3
Mass (earth = 1)	0.0025
Volume (Earth =)	0.01
Mean albedo (geometric)	0.3
Escape velocity	1.2 km/s
Orbital Data	
Mean distance from Sun (10^6) km	5,900.0
Mean distance from Sun (AU)	39.44
Eccentricity of orbit	0.25
Inclination of orbit to ecliptic	17°.1
Orbital period (sidereal)	248 years

CHAPTER 7
THE SUN, A REPRESENTATIVE STAR

The energy that we receive from the Sun dictates the environment on Earth that is so important to humanity's existence. But to astronomers, the Sun is the only star that can be studied in great detail; thus, studying the sun is vital to the understanding of stars as a whole. In turn, the study of stars shows us that our Sun is merely an average star, neither exceptionally bright nor exceptionally faint. Evidence from other stars has also revealed their life histories, allowing us a better understanding of the part and future of our particular star.

Global Properties

The solar diameter equals 109 Earth diameters, or 1,390,000 kilometers. What we see when we look at the sun, however, is not a solid, luminous surface, but a spherical layer, called the **photosphere**, from which the bulk of the solar light comes (see Figure 7-1). Above the photosphere the **solar atmosphere** is transparent, allowing light to escape. Below the photosphere, the physical conditions of the material of the **solar interior** prevent light from escaping. As a result, we cannot observe this interior region from the outside. The solar mass is equivalent to 330,000 earth masses, or 2×10^{30} kg, for a mean or average density (mass/volume) of 1.4 g/cm^3.

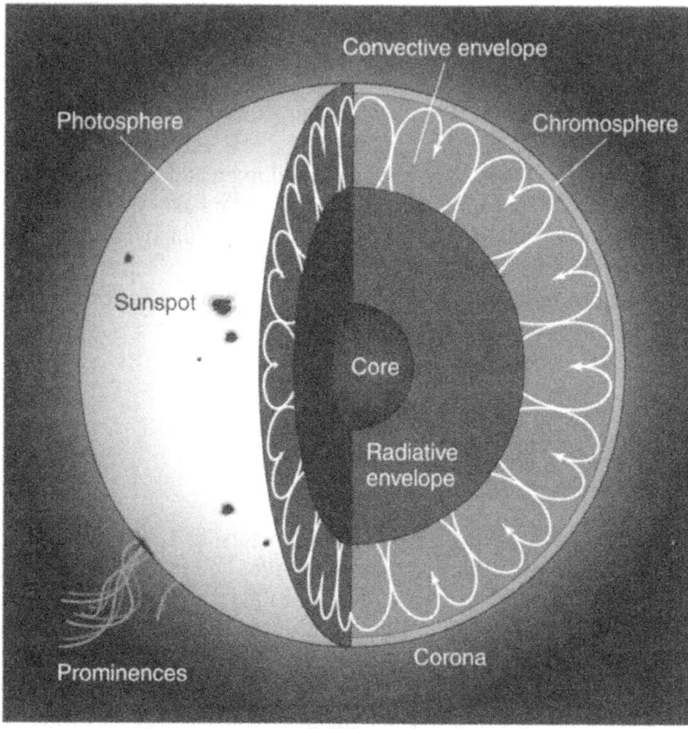

Figure 7-1

Cross-section of the Sun.

The rotation of the sun is made evident by the sunspots that cross the solar disk in about two weeks, then disappear, and then reappear at the opposite limb (or curved edge) two weeks later. Observations of the sun reveal that different parts of the Sun rotate at different speeds. For example, the equatorial rotational period is 25.38 days, but at latitude 35°, the period is 27 days. Sunspots aren't seen at higher latitudes, but use of the Doppler effect for light observed at latitude 75° reveals a longer period of 33 days. This **differential rotation** reveals that the Sun is not solid, but is gaseous or liquid.

The total energy emission of the sun, or **luminosity**, is 4×10^{26} watts. This is found by measurement of the **solar constant**, the energy received per square meter (1,360 watts/m^2) by a surface perpendicular to the direction of the Sun at a distance of 1 astronomical unit and multiplying by the surface area of a sphere of radius 1 AU. The term *solar constant* implies a belief in a constant luminosity output for the Sun, but this may not be completely correct. The **Maunder minimum**, an era of very few detectable sunspots in the century after their discovery in 1610, suggests the solar sunspot cycle (see "The Sunspot Cycle" section later in this chapter) was not in operation at this time. Other evidence suggests the presence or lack of a solar cycle is related to changes in the solar luminosity output. Past ice ages of the Earth could be the result of a diminished solar luminosity output. Monitoring of the solar constant in the last decade from spacecraft suggests there are variations on the order of one-half percent. Thus our Sun perhaps is not as constant a source of energy as was once believed.

The temperature of the solar "surface" (the photosphere) can be defined in several ways. Application of the Stefan-Boltzman Law (energy emitted per second per unit area = σT^4) yields a value of 5,800 K. Wien's law, which relates the peak intensity in the spectrum to the temperature of the emitting material yields T = 6,350 K. This discrepancy between the two values results for two reasons. First, the emitted light comes from different depths in the photosphere and thus is a mixture of emission characteristics of a range of temperatures; thus, the solar spectrum is not an ideal black body spectrum (see Chapter 2). Second, absorption features significantly alter the spectrum from the shape of a black body spectrum.

The strongest absorption features were first studied by Fraunhofer (1814) and are called **Fraunhofer lines**. Absorption lines from over 60 elements have been identified in the solar spectrum. Analysis of their strengths gives temperatures at different depths in the photosphere and chemical abundance ratios. The most common elements are listed in Table 7-1.

Table 7-1: Chemical Composition of the Sun

Element	*Relative Number of Atoms*	*Element*	*Relative Number of Atoms*
H	1,000,000	Ne	37
He	63,000	Fe	32
O	690	S	16
C	420	Al	3
N	87	Ca	2
Si	45	Na	2
Mg	40	Ni	2

Table 7-2 lists the Sun's physical data.

Table 7-2: Physical Data of the Sun

Description of Measurement	*Data*
Diameter	1,392,539 km
Inclination of equator to ecliptic	7°.25
Mean axial rotation period (sidereal)	25.38 d
Mean density	1.41 g/cm^3
Mass	1.989×10^{30} kg
Luminosity	3.85×10^{26} W
Volume (Earth = 1)	1.3×10^6
Escape velocity	617.3 km/s

The Photosphere

The **photosphere** is the visible "surface" of the Sun, but is not a true or solid surface because the Sun is completely gaseous. Moving outward from the core of the Sun, the density, temperature, and gas pressure all decrease until, in a thin layer (only 400 kilometers thick), the material gradually changes from being completely opaque (light cannot pass through it) to being completely transparent.

Close inspection of the photosphere shows that it is not a uniformly bright layer, but has variations. It has a mottled or granular appearance, **granulation,** due to small regions of warmer and cooler material. The brighter regions are about 100 K hotter than surrounding regions and are rising at 2-3 km/s. These regions last only a few minutes before they dissipate and are replaced with other rising masses of warmer gas. Meanwhile, the surrounding cooler and darker regions are sinking back into the solar interior. This is direct observational evidence that the photosphere is also the top of a **convective zone** in the Sun (like the mass motion in a pot of boiling water that carries heat from the bottom to the top of the water). In the deeper interior, photons are able to move energy outward by slow diffusion in a **radiative zone.** Convection is far more efficient at moving energy, with energy passing outward through the solar convective zone (the outer 15% of the Sun) in a matter of hours. Photons, however, endure enormous numbers of collisions with nuclei and electrons and may take tens of thousands of years to work their way into the outer regions of the Sun. Larger areas of the photosphere may appear slightly brighter (warmer) or fainter (cooler) than other regions. This **supergranulation** indicates the existence of larger convective regions.

Within this layer also occur the **sunspots,** the most visible of photospheric features. Sunspots are temporarily cooler regions appearing as dark spots on the surface of the Sun. Sunspots appear dark only in contrast to surrounding hotter regions. In their centers, the temperature is lower by about 1,200 K, and hence their surface brightness is only one-third that of nearby regions. A typical sunspot appears to have a dark center (**umbra**) and lighter outer annulus (**penumbra**) with radial striations caused by the magnetic field associated with the sunspot material. Direct measurement of surface magnetic fields shows that sunspot magnetism is stronger than the surroundings. Sunspots tend to appear in groups (up to 200,000 kilometers in diameter), usually developing rapidly in pairs, then fading away slowly over a few weeks. The leading sunspot in a pair has a magnetic polarity opposite to the trailing spot, with Southern Hemisphere pairs showing the north-south polarity opposite to that of the Northern Hemisphere pairs.

Convection is a turbulent process and produces **faculae,** short-lived hot jets of material that spurt outward into the layers outside the photosphere. **Flares** and eruptions, or brightenings of large regions lasting from minutes to hours, are associated with sunspots and again are a result of the convection. Such flares spew out large quantities of charged particles which, if they encounter the Earth four days later, cause distortions in the Earth's magnetic field (a **geomagnetic disturbance**).

The Chromosphere

The **chromosphere** is the layer (approximately 4,000 kilometers deep) immediately outside the photosphere (from the Greek *chroma* meaning color, a reference to its pinkish appearance due to its primary emission at the red hydrogen wavelength, the Hα line). This region is seen during solar eclipses when light from the brighter photosphere is blocked from view. This region is also known as the **reversing layer** because it is the source of most of the absorption lines seen in the spectrum (whereas the primary process in the underlying photosphere is emission of light). At its base, the temperature is 4,500 K, but at its top, the temperature is 10,000 K, a reversal of the trend for temperature to decrease from the center of the Sun up through the photosphere (see Figure 7-2). Shock waves moving faster than the speed of sound, generated by the convective turbulence in the photosphere, move into the thinner chromosphere, heating its material. Features associated with the chromosphere are the **spicules,** narrow jet-like fountains rising higher into the solar atmosphere. They are perhaps associated with the granules of the photosphere.

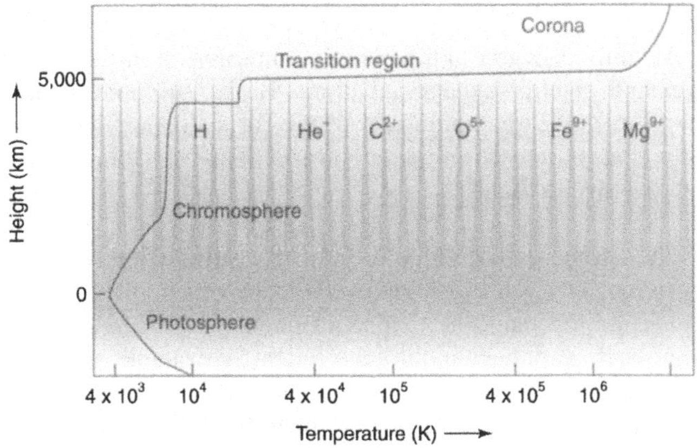

Figure 7-2

Temperature in the photosphere and chromosphere.

The Corona

The **corona** is the very tenuous and very hot, outer atmosphere of the Sun, with temperatures up to 1 to 2,000,000 K. Its light emission is very faint—about as bright as the full moon or one millionth as bright as the solar disk. Therefore the corona is visible only during eclipses (natural or artificial). Magnetic phenomena are apparently responsible for conveying energy from the turbulent photosphere into the material of the corona. This energy shared amongst the very few atoms in the corona produces the high temperatures.

Prominences (when seen near the solar limb) and **plages** (when seen superimposed on the solar disk) are corona regions that appear very bright in the visible part of the spectrum. These features often appear as streamers or filaments, suggesting a structure related to magnetic fields. The conditions of these regions produce excess emission of light and do not necessarily indicate a flow of matter in the corona.

At million degree temperatures, emission of extremely short wavelength electromagnetic radiation—Xrays—becomes important. Viewed in Xrays, the luminosity of the corona is far from uniform, with regions of intense emission (**coronal loops**) and other areas that are dark (**coronal holes**).

The corona grades smoothly into the **solar wind**, an outward flow of ionized gas that has achieved escape velocity. The solar wind bypasses Earth's magnetic field at a velocity of 400 km/s and causes a bow shock; variations in the solar wind (for example, excess particles ejected by flares) may disturb Earth's global magnetic field and can disrupt long-distance radio communication. Solar wind particles trapped by the Earth's magnetic field are funneled into the polar regions producing the aurora. **Airglow** is also a result of solar wind exciting atmospheric molecules to emit light. The night sky emits twice as much light as is received from stars.

The Sunspot Cycle

Sunspots also show a **sunspot cycle** where the number of sunspot occurrences varies between a high value at **solar maximum** and a small value at **solar minimum,** with an approximate 11-year periodicity. Other solar activity (flares, solar wind variations) follows this cycle between an **active Sun** and a **quiet Sun.** Theory suggests that the Sun generates its magnetic field by interior electrical currents in a **solar dynamo.** The differential rotation proceeds to wrap the magnetic field around the Sun and large-scale convection pulls magnetic field lines up and down through the photospheric layer, producing sunspots where compressed field lines move out or into the photospheric layer. After 11 years, more or less, field lines are so jumbled together that the magnetic field disappears, to be regenerated for the next cycle with an opposite polarity. This reversal is associated with a reversal of the polarity of sunspot pairs, hence two sunspot cycles actually are one manifestation of a longer 22-year **solar magnetic cycle.** The solar cycle is also associated with changes in the coronal structure, which appears round at sunspot maximum, but greatly distorted at sunspot minimum.

Internal Structure — The Standard Solar Model

Because light emitted in the interior regions of the Sun cannot be observed, the interior structure of the Sun must be deduced from theory. The **interior structure** is defined by numerical functions that show how every relevant physical factor changes as the radius r increases from r = 0 km at the center of the Sun outward to the radius of the photosphere (r = 700,000 km). The physical factors include mass M(r), density ρ(r), pressure P(r), luminosity L(r), temperature T(r), energy generation rate per unit mass ρ(r), opacity κ(r), chemical composition [the fraction by mass that is hydrogen X(r); the fraction by mass that is helium Y(r); and the fraction by mass that represents all heavier elements Z(r)], and the mean molecular weight μ(r).

Computer calculation of these functions treats the interior of the Sun as if it were composed of spherical layers like the inside of an onion, with conditions slowly changing from layer to layer. The laws of physics relate each layer to the others, providing the mathematical equations that allow each physical quantity to be numerically determined in each layer. These laws include **mass continuity,** which states that in each layer, the addition of mass to M(r) is equal to the density times the surface area of the layer times its thickness. The principle of **hydrostatic equilibrium** states that gas pressure (force per unit area) in each layer must balance the inward gravitational pull or weight of all overlying layers. **Thermal equilibrium** relates the change of energy per second flowing outward through each layer (that is, the luminosity) to the energy generation rate in that layer. The **equation of state** prescribes the relation of gas pressure to the temperature and particle density at any point. Furthermore, in each layer, the computations must check to see how energy is flowing through that layer, by the diffusion outward of photons (radiation) or by mass motion (convection); if the change in temperature over a distance is too great, then photons are unable to carry away energy and hotter material will move upwards into cooler regions (convection). Additional equations allow calculation of the **opacity,** a measure of how opaque the material is. Finally, there are the equations to determine energy generation, which depends on the density, temperature, and chemical composition.

Modern computer programs involve up to 250,000 lines of computer code to obtain a star's interior structure. The results are only weakly dependent upon some necessary assumptions that must be made in these calculations, hence the interior of the Sun is believed to be fairly accurately known and calculations are referred to as the **Standard Solar Model.** In this model, the central conditions are computed to be a density of 150 g/cm^3 and a temperature of 15,000,000 K.

Energy Generation — The Proton-Proton Cycle

The energy radiated away from the solar photosphere is generated in the solar interior by **thermonuclear reactions** involving the fusion of four nuclei of hydrogen to one nucleus of helium. Temperatures are high enough for this to occur only in the central 25 percent of the Sun, called the **core.**

The relevant nuclear reactions are governed by only four physical principles: **conservation of electrical charge** (the net electrical charge does not change in a reaction); **conservation of leptons** (leptons are lightweight nuclear particles such as electrons e^-, positrons e^+, and neutrinos ν); **conservation of baryons** (baryons are heavy nuclear particles such as protons and neutrons, also called **nucleons**); and **conservation of mass-energy** (mass m and energy E are equivalent forms, related by Einstein's equation $E = mc^2$ where c is the speed of light). The specific process that occurs in the central region of the sun begins with the combination of two hydrogen nuclei or protons, hence is termed the **proton-proton cycle.** To keep track of the various particles that make up the nuclei that are involved in these reactions, a notation such as ^{m}X may be used, where m represents the total number of particles (neutrons plus protons) in the nucleus, and X is the chemical species of the nucleus, equivalent to specifying the number of protons in the nucleus. ^{1}H therefore is the normal form of hydrogen, which consists of a single proton; ^{2}H is a heavier form of hydrogen, deuterium, which contains a neutron in addition to the proton; ^{4}He the common form of helium with two protons and two neutrons; and so forth. The basic proton-proton cycle is the sequence:

$$^{1}H + {}^{1}H \rightarrow {}^{2}H + e^{+} + \nu$$

$$^{2}H + {}^{1}H \rightarrow {}^{3}He + \text{energy}$$

$$^{3}He + {}^{3}He \rightarrow {}^{4}He + {}^{1}H + {}^{1}H + \text{energy}$$

For each kilogram of hydrogen, about 0.007 kg disappears into energy via $E = mc^2$ because helium has less mass than four individual hydrogen atoms. To account for the solar luminosity thus requires about 4×10^{38} reactions each second; in other words, the conversion of approximately 6×10^{11} kilograms of hydrogen to helium each second.

At the temperatures that prevail in the solar core, the Standard Solar Model predicts that about 8 percent of the time, the last step in the above sequence is replaced by

$^{3}He + ^{4}He \rightarrow ^{7}Be + \text{energy}$

$^{7}Be + e^{-} \rightarrow ^{7}Li + \nu$

$^{7}Li + ^{1}H \rightarrow ^{4}He + ^{4}He + \text{energy}$

and about 1 percent of the time, another alternative occurs

$^{7}Be + ^{1}H \rightarrow ^{8}B + \text{energy}$

$^{8}B \rightarrow ^{8}Be + e^{+} + \nu$

$^{8}Be \rightarrow ^{4}He + ^{4}He + \text{energy}$

Solar Neutrino Problem

Our understanding of the central region of the Sun is based solely on theoretical calculations. In searching for a means of testing whether these calculations were correct, scientists realized that the neutrinos coming from the nuclear reactions would allow such a test. **Neutrinos** are very elusive particles. Originally believed to be massless, they travel essentially at the speed of light and have an extremely low likelihood of ever reacting again with other forms of matter. The boron to beryllium reaction in the proton-proton cycle, however, has a special significance for scientists on Earth; this neutrino is a high-energy neutrino that has the greatest possibility of being experimentally detected at Earth via the reaction

$^{37}Cl + \nu \rightarrow ^{37}A + e^{-}$

Measurement of the number of these neutrinos arriving at Earth then allows deduction of the number of reactions occurring in the core of the Sun. Because the reaction rate is very temperature dependent, astronomers have, in essence, a means of determining the true central temperature of the Sun to compare with the prediction of the Standard Solar Model. In an experiment begun in the 1960s, 100,000 gallons of the cleaning fluid perchloroethylene (C_2Cl_4; containing 2×10^{30} atoms of ^{37}Cl) has been used as the neutrino detector: a few chlorine atoms in the molecules of the cleaning fluid interact with solar neutrinos to produce argon. One then recovers the argon atoms from the tank to determine how many argon atoms are produced in each time period. The expected capture rate of neutrinos for the Standard Solar Model is 1 neutrino per day! After 30 years of operation, the actual capture rate is only one-third to one-half of this, a result that if correct could actually rule out the proton-proton cycle as the source of the Sun's energy. Newer measurements of solar neutrinos, in a variety of other experiments that also detect other neutrinos from the thermonuclear reactions, have confirmed the apparent deficiency of the solar neutrino flux. But three decades of theoretical effort have found no satisfactory modification of the Standard Solar Model that can account for a lower number of neutrinos.

At present, it appears that the problem has not to do with understanding of the Sun but rather with the physics of neutrinos. Neutrinos exist in three forms (each associated with a specific of nuclear particle): **electrons, muons,** and **tauons.** If neutrinos actually have a tiny amount of mass (the original theory predicted that they were massless), then solar neutrinos can change. They leave the Sun as electron neutrinos, but in the eight minutes of travel between the Sun and Earth, they become equal numbers of the three neutrino types. Additional experiments are underway to test the hypothesis that neutrinos do have mass.

Helioseismology

In recent years, another method of study, **helioseismology,** has been developed to determine internal conditions in the Sun. The Sun is vibrating with a complex spectrum of frequencies, due to the propagation of resonant sound waves through its outer layers. These

solar oscillations are the product of the Sun's internal structure in the same fashion that the vibrations of a drum are related to the drum's physical structure. Helioseismology has confirmed predictions of the depth of the outer convective layer made by the Standard Solar Model.

CHAPTER 8
OBSERVATIONAL PROPERTIES OF STARS

Direct astronomical investigation of the properties of stars relies principally upon only three types of study. **Astrometry,** the measurement of stellar positions, yields parallaxes, proper motions, and apparent orbital motions (of binary stars). **Photometry,** the measurement of the quantity of light received at all wavelengths or within specific wavelength regions, gives apparent brightnesses (apparent magnitudes), colors, and variability of stellar brightnesses. **Spectroscopy,** the dispersion of light into its component colors to determine the stellar energy distribution as a function of wavelength, produces spectral types, luminosity classes, strengths of absorption and emission features from which chemical composition may be deduced, and Doppler shifts. More specialized interference techniques applied to starlight can produce data on stellar angular sizes and some idea of differences in the brightness across the photospheres of distant giant stars.

Understanding the physical nature of stars relies upon only a small number of properties that can be directly inferred from these three observational techniques. The first observational factor is distance (although distance is not really a *property* of a star, in the same sense that a person is the same person whether here or far away). The knowledge of a star's distance is needed to reduce observed factors (apparent brightness, apparent size, and so on) to absolute quantities (absolute brightness or luminosity, radius, and so on).

Stellar Parallax and Distances

For nearby stars, distance is determined directly from parallax by using trigonometry and the size of Earth's orbit. The **trigonometric** or **stellar parallax** angle equals one-half the angle defined by a baseline that is the diameter of Earth's orbit. Because even the nearest stars are extremely distant, the parallax triangle is long and skinny (see Figure 8-1).

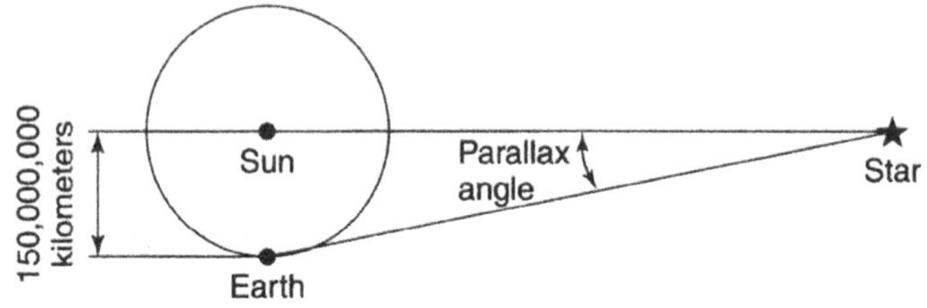

Figure 8-1

Parallax.

The relationship between the parallax angle p" (measured in seconds of arc) and the distance d is given by d = 206,264 AU/p"; for a parallax triangle with p" = 1", the distance to the star would correspond to 206,264 AU. By convention, astronomers have chosen to define a unit of distance, the **parsec,** equivalent to 206,264 AU. The parsec, therefore, is the distance to a star if the parallax angle is one second of arc, and the parallax relation becomes the much simpler form

d (in parsecs) = 1/parallax angle in seconds of arc

A more familiar unit of distance is the **light-year,** the distance that light travels (c = 300,000 km/s) in a year (3.16×10^7 seconds); one parsec is the same as 3.26 light-years.

The nearest star, α Centauri, has a parallax angle of 0.76". Therefore its distance is d = 1/0.76" = 1.3 pc (4 ly). The ground-based limit of parallax measurement accuracy is approximately 0.02 arc second, limiting determination of accurate distances to stars within 50 pc (160 ly). The European Hipparcos satellite, in orbit above the atmosphere and its blurring effects, can make measurements with much higher precision, allowing accurate distance determinations to about 1000 pc (3200 ly).

Apparent Magnitudes

Apparent magnitude (for which the symbol m is used) is a measure of how bright a star looks to the observer. In other words, it is a measure of a star's **energy flux,** the energy received per second per square meter at the position of the observer. The magnitude scale was created by Hipparchus, who grouped the stars he could see into six categories or **magnitudes.** Magnitude 1 stars are the brightest naked-eye stars, magnitude 2 stars are fainter, magnitude 3 stars even fainter, and so forth, down to magnitude 6, the faintest stars visible to the naked-eye. When it became possible to make reliable energy measurements of starlight, it was found that a *difference* of 1 magnitude corresponds to a *ratio* of brightnesses (or fluxes) of about 2.5 times. The modern magnitude scale makes this quantitatively very precise: By definition, a difference of 5 magnitudes corresponds to a brightness (flux) ratio of 100. In turn, a difference of 2.5 magnitudes corresponds to a brightness ratio of $\sqrt{100} = 10$, and a difference of 1 magnitude corresponds to a brightness ratio of $\sqrt[5]{100} = 2.512$, and so forth. Magnitudes m therefore represent a logarithmic scale; for example, for two objects 1 and 2 with respective magnitude m_1 and m_2,

$$m_1 - m_2 = 2.5 \log(f_2/f_1)$$

where the brightness or flux (f) is measured in watts/m^2. Flux is related to both the intrinsic energy output of an object (its **luminosity** L) and the area ($4\pi r^2$) over which that energy has spread as it moves out to a distance r from the object:

$$f = L/(4\pi r^2)$$

Therefore, an apparent magnitude depends on both the true luminosity and the distance of an object.

The original magnitude scale was limited to naked-eye stars, but it can be applied to even brighter objects by extending to negative numbers (see Figure 8-2). The apparent magnitude of the Sun is –26.8 and the full Moon –12.5. Fainter objects have larger positive numbers.

The Hubble Space Telescope can detect objects as faint as m = +30 (24 magnitudes, or about four billion times fainter than the faintest naked-eye stars).

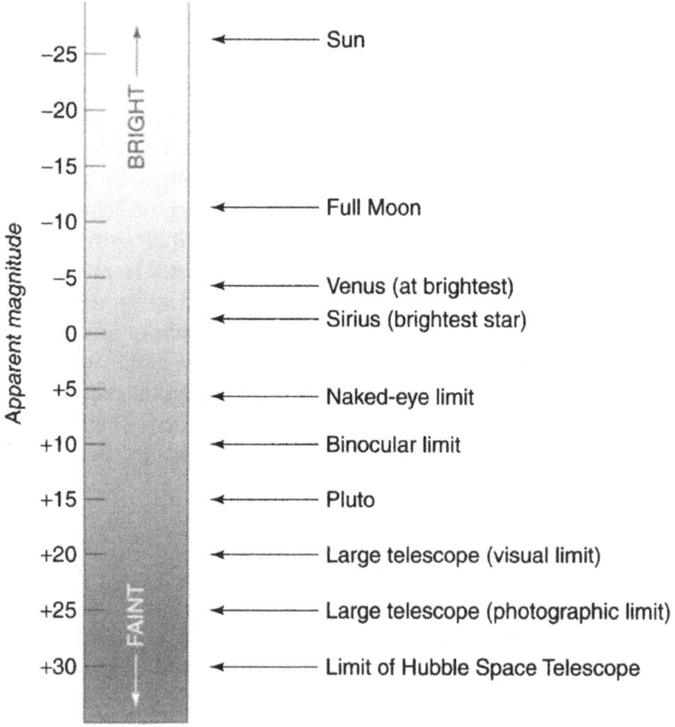

Figure 8-2

Apparent magnitude scale.

Originally, magnitudes were based on visible light; but with the use of filters, astronomers can measure brightnesses in specific parts of the spectrum. A U magnitude is one measured using the ultraviolet light of a star, B represents a blue magnitude, V a visual (yellow-green part of the spectrum) magnitude, with R, I, J, K, M, and so on, representing ever longer wavelengths in which the measurement is made.

Absolute Magnitudes

To use the magnitude system to compare stars on an absolute scale means removal of the distance effect. An **absolute magnitude** (for which astronomers use the symbol M) is defined as the apparent magnitude a star would have if it were placed at a standard distance of 10 parsecs (32.6 light-years). It can be shown that the *difference* between apparent and absolute magnitude is related to the distance as

$$m - M = 5 \log (r/10 \text{ pc}) = 5 \log r - 5$$

where distance r is expressed in parsecs. This difference m – M is called the **distance modulus**. $m - M = -5$ corresponds to a distance $r = 1$ pc, $m - M = 0$ corresponds to $r = 10$ pc, $m - M = +5$ corresponds to $r = 100$ pc, and so on. If the Sun were moved to a distance of 10 pc, it would appear as a 5th magnitude star, that is, $M \sim +5$ (it is useful to remember this, that a star of 1 solar luminosity corresponds to an absolute magnitude of +5).

Luminosities

Luminosity is a direct measure of the total energy radiated away by a star, expressed in joules/second or watts. If one has the flux and the distance, then one can calculate the luminosity of a star. That is:

$$L = 4\pi r^2 f$$

which is usually expressed in units of the solar luminosity (4×10^{26} watts = 1 $L_\odot$). Luminosity (sometimes called **absolute brightness**) and absolute magnitude are two different ways of expressing a measurement of the same aspect of a star. From the definition of magnitude, therefore, one has

$M = -10$ corresponding to $L/L_{\odot} = 10^6$;

$M = -5$ corresponding to $L/L_{\odot} = 10^4$;

$M = 0$ corresponding to $L/L_{\odot} = 10^2$;

$M = +5$ corresponding to $L/L_{\odot} = 1$;

$M = +10$ corresponding to $L/L_{\odot} = 10^{-2}$;

$M = +15$ corresponding to $L/L_{\odot} = 10^{-4}$;

and $M = +20$ corresponding to $L/L_{\odot} = 10^{-6}$.

The span of absolute magnitude $-10 < M < +20$ covers the full range of absolute stellar brightnesses for normal stars.

Masses

Mass is the most fundamental property of a star. In fact, the amount of total mass in a star and the way that mass is distributed among the various chemical elements essentially determines all the other properties of the star, including its size, luminosity, and surface temperature. But unlike size (which can be found from a star's apparent angular diameter and distance), luminosity (which can be calculated from the star's apparent brightness and its distance), and surface temperature (which can be deduced from spectroscopy), there is no easy means of measuring the mass of most stars. Mass determination can be accomplished only indirectly, by observing separations between objects and their motions; motions are changed by their mutual gravitational attraction, which depends on both masses and distances.

Thus, knowledge of two factors (motion and separation) yields the desired mass. The simplest circumstance is if two stars are in orbit about each other forming a binary star system, then Kepler's Third Law (see Chapter 2) may be applied to obtain the stellar masses from the size and period of the orbit. Analysis of orbital motion even today has yielded accurate masses for only a limited number (less than 100) of stars.

Radii

The determination of the radius of a star requires observation of its angular size, which with its distance yields the star's true linear size. However, because stars are so distant, they appear to be only a fraction of an arc second across, an apparent diameter that is vastly increased by the blurring effects of the atmosphere. The apparent diameter can also be increased by the diffraction effects in a telescope that is in orbit above the atmosphere. There are sophisticated interferometric techniques to measure these tiny angular sizes, but they are difficult to put into practice and not feasible for obtaining large quantities of data. One of these methods is **speckle interferometry,** a technique in which many short exposure images of a star are recorded at a telescope. The exposure time must be a fraction of a second in order to prevent atmospheric motions from blurring the image, but then each image is insufficiently exposed to reveal any significant detail. Hundreds of images, however, can be combined to form an image with sufficient resolution to determine the angular size of a star and even some detail of the surface features in the case of supergiant stars.

On the other hand, sizes of stars may be more easily obtained by observation of the light coming from eclipsing binaries. If the orbital properties are known, then the length of an eclipse directly gives the size of the eclipsing and eclipsed stars.

More generally, a stellar radius R can be directly inferred from use of Stefan-Boltzman's Law (see Chapter 2) if the surface temperature T and luminosity L are known for a star, that is, from the relationship

$$L = 4\pi r^2 \sigma T^4$$

Colors

A **stellar color** may be numerically defined by subtracting the magnitudes of a star through a given pair of filters such as B from U, or V from B, for example U–B and B–V, yielding magnitude quantities that represent a measurement of the ratio of the ultraviolet to blue light and the blue to the visible light, respectively. Because magnitudes run backwards, a large B-V means B>V, so the star is actually fainter in B than V, so it is a red star. So these values are indeed related to the color of the star with numerically large positive values (1.6–2.0) of U–B and B–V obtained for red (cool) stars, but values near 0.0 or even negative indicate blue (hot) stars. Photometric colors often are used instead of surface temperatures or spectral types for stars.

Spectral Types

The majority of the light emitted by a star comes out of its dense interior and thus shows a characteristic black body or continuous spectrum of energy distribution. Superimposed upon this continuous spectrum are the absorption lines due to absorption at specific wavelengths by atoms in the cooler atmosphere immediately above the stellar photosphere. The specific pattern of absorption features is related to the chemical composition of the absorbing material. However, because most stars are chemically similar, the more important factor in determining which absorption features are actually present in a spectrum is the photospheric temperature of a star.

The system of spectral classification was designed by Annie Jump Cannon at Harvard University. On the basis of the number and pattern of absorption features, she classified stellar spectra into classes A (simplest looking), B (next simplest), and so on, up to type W (most complex). Years later when the newly developed understanding of atomic physics was applied to interpretation of stellar spectra, scientists realized that temperature was the key factor that determined at what wavelengths a chemical element present in a stellar atmosphere would absorb light (see Figure 8-3).

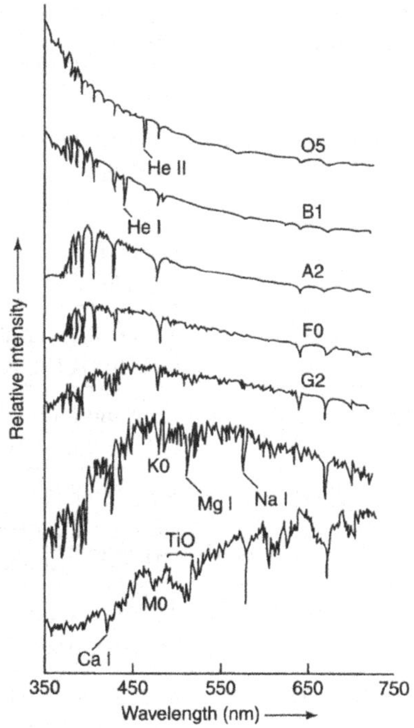

Figure 8-3

Illustration of spectral types. Some of the elements responsible for the absorption features are marked (Ca calcium; He heliem, Mg magnesium, Na sodium, and TiO for titanium oxide).

When ordered in terms of decreasing photospheric temperature, the surviving spectral types form a sequence **O B A F G K M**. For historical reasons, O and B stars are often referred to as early-type stars, and K and M stars are known as late-type stars. With refinement of the system of spectral classification, each spectral type was further broken down in tenths of classes, for example A0, A1, . . . , A9, F0, and so on. The spectral types directly correlate with temperature, as shown in Table 8-1 and Figure 8-4.

Table 8-1: Spectral Types, Temperatures, and Absorption Features

Type	*Representative Temperature*	*Strongest Spectral Lines Produced by These Elements*
O	28,000 K	He, He^+, C^{++}, H
B	>28,000 K (B0)	He (strongest at B2, disappears by B9), C^+, H strength increases from B0 to B9
A	10,000 K (A0)	H maximum absorption at A2, Ca^+, Fe^{++}
F	7,600 K (F0)	H, but weaker absorption than for A stars
G	6,000 K (G0)	Prominent metal lines (Ca^+, Fe^+, Fe); also H
K	4,800 K (K0)	Metal lines (Ca^+, Fe^+, Fe) stronger than H;weak CN, CH, other molecules
M	3,300 K (M0)	TiO molecular bands increasing (maximum at M7);
	2,200 K (M5)	VO in coolest stars, also Ca, Fe

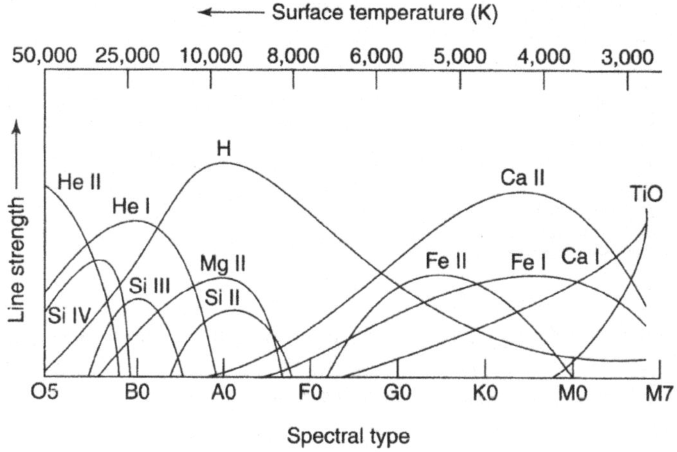

Figure 8-4

Relative strength of absorption lines as a function of temperature and spectral type.

In addition to the OBAFGKM types, astronomers still use spectral types W, R, N, and S. Type W stars are variants of type O stars, the central stars of planetary nebulae, which are very hot and often show emission lines. R and N stars have the same temperatures as K and M stars, respectively, but show extremely strong carbon and carbon molecule absorption features (these are also called **carbon stars**). S stars, also similar to M stars in temperature, have a peculiar chemistry and show absorption by zirconium oxide (ZrO) and lanthanum oxide (LaO). Recently spectral types L and T have been proposed for the very coolest and faintest stars that have been discovered. Type L stars have temperatures between 1,300 and 2,000K; their spectra show absorption by iron hydride (FeH) and chromium hydride (CrH) molecules. Even cooler objects, temperatures between 700 and 1,300K and that show strong methane absorption, would be termed T spectral types. These very coolest objects may not be true stars in the sense of being able to generate energy by central thermonuclear reactions; they have also been termed *brown dwarfs.*

A primary advantage of classifying stars by spectral type is that with even a modest amount of experience, a person can easily recognize a star's spectral type from a photographic spectrum or a digital spectrum that shows quantitatively the energy distribution with wavelength.

Surface Temperature

The **surface temperature** (more correctly, photospheric temperature) of a star may be obtained from a quantitative analysis of the absorption lines that appear in a spectrum. Temperature is closely correlated with stellar spectral type (as shown previously in Table 8.1) and with photometric color.

Chemical Composition

Detailed spectral analysis also yields the **chemical composition** of the material (photosphere) that produced the absorption lines in a stellar spectrum. Most stars found in the Galaxy are chemically very similar to the Sun, approximately 74 percent hydrogen, 24 percent helium, and 2 percent all other elements (all elements heavier than helium are conventionally grouped together under the term **metals**). Stars do exist for which the heavy element fraction is larger (**metal-rich** stars) and others for which the heavy elements are deficient (**metal-poor** stars).

Luminosity Classes

The gas pressure in the photospheric layer of a star can also affect the way absorption features appear. If the pressure is low (as would be the case in the atmosphere of a supergiant star due to the large size giving it a low surface gravity), the atoms absorb at the specific wavelengths set by their internal atomic energy levels. But if the pressure is higher (as in a giant star), the higher pressure means that the individual atoms are packed more closely together. The electronic properties of atoms will affect each neighbor's internal energy levels and the atoms will be able to absorb at slightly smaller and slightly greater wavelengths—the absorption features will appear broader. In a small star (like the Sun), the gas pressure must be higher to balance against gravity. The atoms are even more closely packed together, neighboring atom's electronic properties are even more affected, and the absorption features will appear still wider.

A second classification of stellar spectra therefore may be done on the basis of the appearance of the absorption lines, with the narrowest line spectra designated I (subdivided into Ia and Ib) through types II, III, and IV, to V, the broadest line spectra. For stars of a given temperature, narrow lines correlate with low pressure atmospheres, large stellar radii, and hence a high luminosity: Type Ia, therefore, are the brightest supergiant stars, and type Ib are the fainter supergiants. Type III stars are still large stars of intermediate brightness, and are termed giants. Type V stars are like the Sun, dwarf or Main Sequence stars, which are fainter than the larger stars. Because this line-width classification correlates with luminosity, the designation Ia to V is termed a **luminosity class.**

The full spectral classification of a star therefore includes both the spectral type (OBAFGKM) as well as its luminosity class (I-V); for example, we would classify the Sun as a G2 V star.

Proper Motions and Radial Velocities

Radial velocity (measured in km/s) is the velocity along the line of sight away from (considered a positive velocity) or toward (negative velocity) the observer. (Astronomers actually correct observed motions for that of Earth, hence recorded velocities are relative to the Sun.) Radial velocity is determined from the Doppler effect in the spectra of the stars. Of the nearby stars, some are moving toward us and others are moving away—there is no indication of systematic collapse or expansion of the Galaxy.

Proper motion is the rate of angular drift across the sky (measured in arc seconds per year) and is found from the star's change of position on the sky (see Figure 8-5). It is related to a star's **transverse velocity** (km/s; the velocity component of the star parallel to the plane of the sky) depending upon the distance to the star. Proper motions are small; Barnard's star, which has the largest known proper motion of 10.31"/yr, would take 175 years to drift an angular distance equal to the diameter of the full Moon.

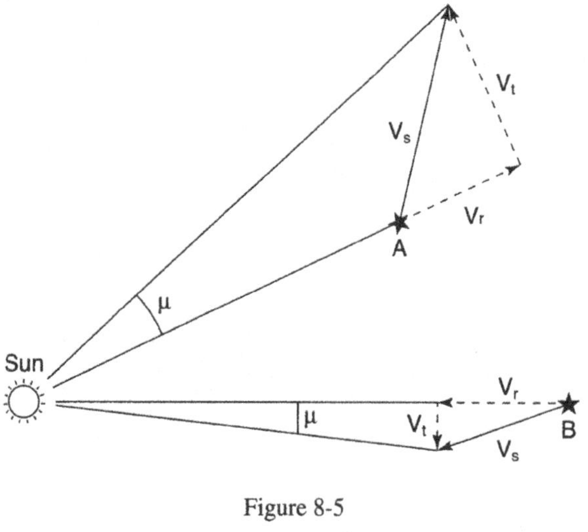

Figure 8-5

Proper motion and space motion. The Doppler effect gives us the radial velocity V_r toward (star B in this example) or away (star A) from the Sun. Each star's proper motion μ combined with its distance yields the transverse velocity V_t. The hypotenuse of the right angle triangle formed by V_r and V_t is the true space motion V_s of each star.

Nearby stars have relatively large proper motions and, in fact, that is how these stars have generally been found. More distant stars have smaller proper motions, hence the distant stars can be thought of as "fixed."

Combining the radial motion toward or away from the Sun with a star's transverse motion at right angles to the line of sight yieldes the star's **space motion** (or true motion), a speed in km/s and a direction relative to the Sun.

Properties of Secondary Importance

A number of other observable factors may be determined for stars, but these generally are of minor importance in understanding stars as a whole. These include evidence for motions such as rotation or turbulence in stellar atmospheres, direct measurement of surface gravity, the presence of circumstellar material, and the strength of a magnetic field. Stars may also exhibit variability in luminosity and size (pulsating and other variables).

Stellar ages should be mentioned because age is a property of any star. But the reality is that age is not determinable directly for most stars and generally must be deduced from theoretical principles applied to understanding how stellar properties change over time. Thus stellar age is not an observational property.

CHAPTER 9
THE HERTZSPRUNG-RUSSELL DIAGRAM

The fundamental tool for presentation of the diversity of stellar types and for understanding the interrelations between the different kinds of stars is the **Hertzsprung-Russell Diagram** (abbreviated HR diagram or HRD), a plot of stellar luminosity or absolute magnitude versus spectral type, stellar surface temperature, or stellar color. The various forms of the HR diagram come from the different manner in which stars may be studied. Theoreticians prefer to graph directly the numerical quantities that come from calculations, for example, luminosity versus surface temperature (see Figure 9-1a). On the other hand, observational astronomers prefer to use those quantities that are observed, for example, absolute magnitude versus color (a photometrist's color-magnitude diagram is essentially the same as an HR diagram) or absolute magnitude versus spectral type (see Figure 9-1b).

The only stars for which absolute magnitude can be directly obtained are the nearby stars for which parallaxes may be measured and hence distances determined; given a distance, an apparent magnitude can be converted to an absolute magnitude. Inspection of a tabulation of stars out to 5 parsecs (16 ly, the distance to which astronomers have a reasonably complete sample of existing stars; at larger distances, there is an increasingly higher probability that the faintest stars have been missed) shows there to be 4 A stars, 2 F, 4 G, 9 K, and 38 M stars. Even these few stars are sufficient to show three general aspects of stars. First, the typical star is much fainter and cooler than the Sun. Second, the fainter the star, the more stars there are. And last, there is a general trend in the sense that the cooler the star, the fainter it is. This track of stars that runs from high luminosity, hot stars to low luminosity, cool stars is known as the **Main Sequence**. A few stars also are found in a clump to the bottom left of the HR diagram, at relatively high surface temperatures, but low luminosities. These stars have been termed **white dwarfs,** and the differentiation of their observational properties from the main sequence stars shows that they must be a very different type of star internally.

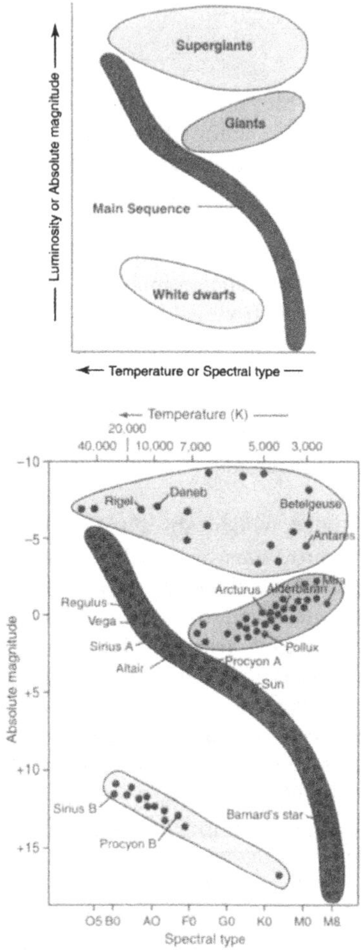

a.

b.

Figures 9-1a and 9-1b

Hertzsprung-Russell Diagrams. Top: The general labeling of stars into four groups is shown. Bottom: Nearby stars and some of the brighter stars in the sky have been added, with the positions of a few well-known stars marked.

The sample of nearby stars contains no highly luminous stars. A survey of greater distances requires the Hipparcos satellite or the application of alternative distance determination techniques, such as those involving star clusters. A cluster of stars may have fainter and brighter stars all at the same distance. Those fainter stars that show a trend from high luminosity, hotter surfaces to low luminosity, cooler surfaces are similar to the main sequence stars in our solar neighborhood. At a given spectral type, those stars must have the same absolute magnitude as the nearby stars, and these absolute magnitudes may be compared with the measured apparent magnitudes to obtain the distance to the cluster. With a known distance, the apparent magnitudes of the brightest stars may also be converted to absolute magnitudes, making it possible to plot these stars in an HR diagram. By use of **main sequence fitting** applied to star clusters (as well as other, more sophisticated techniques), the upper (brighter) portion of the HR diagram may be filled in. Such a technique enhances the importance of the HR diagram — it is not only a means to display (some of) the properties of stars, but it becomes a tool by which information about other stars may be derived. (See Figure 9-2.)

When a large number of stars are plotted in the HR diagram, it becomes clear that the main sequence stars are represented across the full range of spectral types as well as across the full range of absolute magnitudes. The hottest main sequence stars have absolute magnitudes $M \approx -10$ and the coolest $M \approx +20$, and alternatively, luminosities that go from 10^6 to 10^{-6} solar luminosities. The Sun is at the middle point of this luminosity range and, in that sense, could be considered an average star.

In addition to the main sequence stars and the white dwarfs, two other distinct groupings of stars may be noted. The first is a concentration of stars with moderately high luminosities ($M \approx -2$ to -4 or so) and relatively cooler spectral types (to the right) of the main sequence. These stars are called **giants** or **red giants.** The second is a distribution of stars at high luminosities ($M < -5$), thinly scattered across the top of the HR diagram, representing the full range of spectral types from O to M. These stars are called **supergiants.**

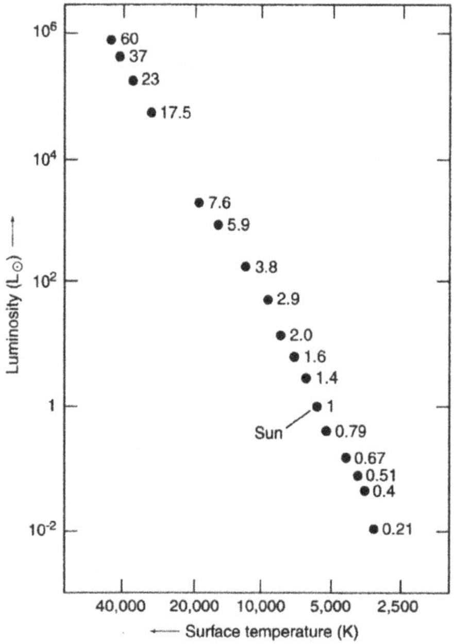

Figure 9-2

Schematic diagram for computed models of main sequence stars, showing luminosities in units of the Sun's luminosity and surface temperature in Kelvins. Adjacent to each model star is its mass in units of the mass of the Sun.

Consideration of the luminosities of the apparent brightest stars in the sky shows they appear bright because they are intrinsically bright. Of these stars, there are only five with $M < -5$ (for example, with luminosity $L > 10^4$ solar luminosities). These are the most luminous stars within a distance of 430 pc, the greatest distance to any of these five (the bright summer sky star Deneb). The volume of space centered on the Sun enclosed by a sphere of this radius is $4\pi(430 \text{ pc})^3/3 = 330{,}000{,}000$ cubic parsecs, yielding an average stellar density of 5 stars / 330,000,000 $\text{pc}^3 = 1.5 \times 10^{-8}$ stars/pc^3. In contrast, there are 38 cool, low luminosity M stars within 5 parsecs of the Sun, in a volume of space $4\pi(5\text{ pc})^3/3 = 520$ cubic parsecs, for an average density of 34 stars / 520 $\text{pc}^3 = 0.065$ stars/pc^3. The ratio of cool main sequence M stars to all classes of highly luminous stars is a factor of 4.4 million. Highly luminous stars are rare, whereas the cool, faint stars are quite common. In this sense, the Sun is actually one of the brighter stars in the Galaxy.

Main Sequence Stars

The fact that the main sequence stars are represented by a band across the HR diagram that is smoothly populated from the rare O and B stars to the very common M stars strongly suggests that these stars are physically the same type of object, though some factor must be responsible for their range in observable properties. The Sun is a main sequence star and thus, by implication, all other main sequence stars must share its fundamental nature. Through theoretical modeling of the Sun and other main sequence stars, scientists have determined that the factor that differentiates them from the three other types of stars is the fact that their energy is generated internally by the conversion of hydrogen to helium (giants and supergiants produce their energy by gravitational contraction and by converting helium to even heavier elements; white dwarfs are like dying embers in a fireplace, radiating away their store of heat energy). Like most other stars, they also are in a state of equilibrium in which gravity is balanced by gas pressure at each radius, and the luminosity flowing outwards at each level is balanced by the energy generated interior to that level.

Mass-Luminosity Relationship

The study of binary stars provides the key piece of information to understanding why main sequence stars have a range of properties from high luminosity to low luminosity. Orbital periods and orbital radii via Kepler's Third Law yields the mass of stars (see Chapter 2). When the luminosity of main sequence stars is plotted against their masses, we observe a **mass-luminosity relationship,** approximately of the form $L \propto M^{3.5}$ (see Figure 9-3). In other words, doubling the mass of a main sequence star produces an increase in luminosity by a factor $2^{3.5} = 11$ times. The Main Sequence is therefore a mass sequence, with low mass stars forming an equilibrium with a cool surface and a low luminosity (low energy generation rate), and high mass stars having hot surfaces and high luminosity (larger energy generation rate). The mass-luminosity relation holds only for main sequence stars. Two giant or supergiant stars with the same luminosities and surface temperatures may have dramatically different masses.

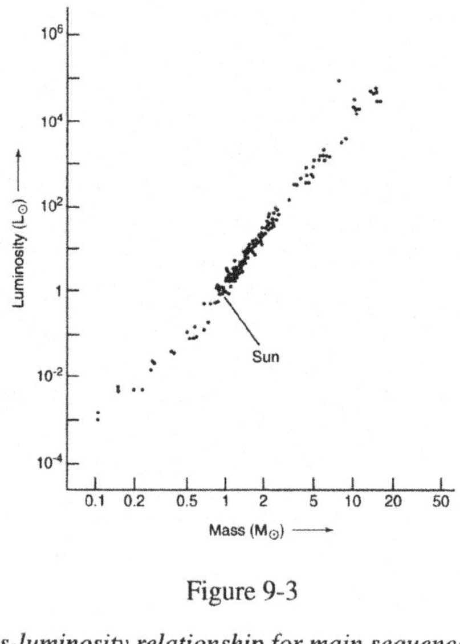

Figure 9-3

Mass-luminosity relationship for main sequence stars.

The fact that luminosity is not directly proportional to mass produces a major problem for observing and interpreting the universe. This problem can be seen by considering the masses, luminosities, and mass-luminosity ratios M/L for different types of stars (only three types are considered in Table 9-1 for simplicity):

Table 9-1: Mass and Luminosity for Main Sequence Stars

Star	*Mass M (solar units)*	*Luminosity L (solar units)*	*M/L (solar units)*
O star	50	10^6	0.00002
G2 star	1	1	1
M0 star	0.1	3×10^{-4}	300

One million G2 stars like the Sun or three billion M0 stars produce the same amount of light as one O star. That one O star represents 50 solar masses of material, whereas the G2 stars would have a total of 1,000,000 solar masses and the M0 stars would have 300,000,000 solar masses. Even though hot, massive, luminous stars are rare, they can easily outshine the vast bulk of the more common stars. In almost any kind of stellar system (star cluster, galaxy), the light is dominated by that coming from only a small fraction of that mass, yet the real structure of any stellar system is represented by the distribution of its mass in space. The visible appearance of an object may actually be rather misleading in terms of what an object actually is.

The main sequence is bounded by upper and lower mass limits of about 80 solar masses and about 0.08 solar masses, respectively. Why should there not be more massive (and hence brighter) and less massive (hence fainter) stars? The upper limit (**Eddington limit**) is set by radiation pressure in the star's photosphere. An 80 solar mass star is not that much bigger than the Sun, but its luminosity is 10^6 times greater. The radiation passing through each square meter of photosphere is perhaps 10^4 times greater than for the Sun. Radiation can apply a pressure (force per unit area) when it interacts with matter because photons of light can act as particles. In collisions with atoms, the atoms can be kicked away from the star. At the upper mass limit of main sequence stars, the addition of a bit more mass would increase the luminosity and radiative flux and simply blow away what has been added. Stable stars in a main sequence state with more than about 80 solar masses simply cannot exist.

The lower mass limit on stars seems to be about 0.08 solar masses. Below this mass limit, internal temperatures and pressures are too low to sustain thermonuclear conversion of hydrogen to helium. Without a thermonuclear energy source, an object is not self-luminous. It would be what has been called a failed star. Such objects actually exist and radiate at infrared wavelengths due to their store of heat energy generated when they contracted gravitationally — these are termed **brown dwarfs.** Less massive objects are planetary bodies like Jupiter.

Red Giants and Supergiants

Two stars at the same spectral type, say type G, can have quite different brightnesses. One could be a main sequence star with $M = +5$ and the other a giant star with $M = -2.5$. By the definition of spectral type, both stars have the same surface temperature T, yet their luminosities L differ by 7.5 magnitudes or a factor of 1000 in luminosity. Stefan-Boltzman's law (see Chapter 2) allows the luminosity of each star to be expressed in terms of its surface temperature and surface area. For example, $L = \sigma T^4 4\pi R^2$, where R is the radius of the star. Relating the luminosity of the first star to the second,

$$L_1/L_2 = (R_1/R_2)^2$$

A ratio of 1000 in luminosity means the more luminous star must be $\sqrt{1000} = 31$ times larger than the main sequence star. As the Sun has a radius of 700,000 km, the radius of the more luminous star is 22 million km. If such a star were placed in the center of the solar system, its surface would be one-third of the distance out to the planet Mercury. As seen from Earth, it would appear 15° in diameter. It is because of the size of these objects that they are termed giant stars. As most stars of this type are cooler and redder, the term **red giant** is often used.

A similar comparison with an even brighter G star at $M = -7.5$ produces a size that is 310 times that of the Sun, or a radius of 220 million km, placing the photosphere at the orbit of Mars if this star were to replace the Sun in the solar system. These immense stars are accordingly termed **supergiants.**

The classification as a giant or supergiant star, however, is as much dependent upon the grouping of stars in the HR diagram as upon radial size. There are giant stars that are actually larger than some supergiant stars.

White Dwarf Stars

The brightest naked-eye star is Sirius. Sirius is actually a binary star system with the two components designated Sirius A and Sirius B. Sirius A is an A star and has the same color, hence same surface temperature as its companion; but Sirius B is about 10 magnitudes (or 10,000 times) fainter than A. The two stars therefore must differ in size (using the same argument as for giant and supergiant stars) by a factor $\sqrt{10{,}000} = 100$ times. Because Sirius A is a main sequence star approximately the same size as the Sun, Sirius B is only slightly larger than Earth. The binary orbit of this pair allows determination of their masses. Sirius A has a mass of twice that of the Sun and Sirius B is 1 solar mass. One solar mass in a volume the size of Earth represents an average density of 2,000,000 grams per cubic centimeter — a spoonful of material from Sirius B would weigh (on Earth) two metric tons! Because of their size and color (these hot stars produce more blue than red light, but the spectral sensitivity of the human eye makes them appear white), these compact stars are therefore termed **white dwarf stars.**

Spectroscopic Parallax

The HR diagram becomes immensely valuable as a tool to determine distances to far away stars when the additional information provided by classification of spectral line widths is considered. Luminosity classes Ia through V have been calibrated as a function of spectral type and may be added to the HR diagram (see Figure 9-4).

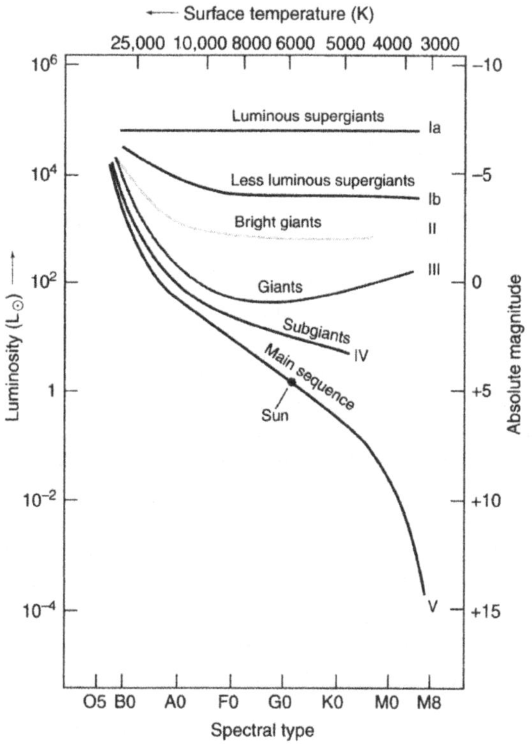

Figure 9-4

HR diagram with luminosity classes added.

From the spectrum of a star, astronomers can obtain both the spectral type (OBAFGKM) and luminosity class (Ia–V). Therefore, where the star is placed in the HR diagram is known, but most importantly, the absolute magnitude of the star can be read off the vertical scale on the diagram. A comparison of the absolute brightness with the star's measured apparent brightness gives the distance to the star. This technique of **spectroscopic parallax** will give the distance to any star for which a spectrum can be obtained. There are other techniques for getting distances to very distant objects, but this is the most important one and permits distance determination even into nearby galaxies several million parsecs away.

CHAPTER 10
THE STRUCTURE OF STARS

The internal structure of normal stars is fairly simple because only a few physical principles are involved in the determination of the structure of a gaseous object. This simplicity is summed up in a simple principle, the **Russel-Vogt theorem.** For stars of like composition, the structure and observable properties depend on a single parameter such as the star's mass. Alternatively, the Russell-Vogt theorem may be expressed as follows: The equilibrium structure of an ordinary star is determined uniquely by its mass and chemical composition. This principle or theorem is not an accident of nature, but is the direct result of how the laws of physics determine the equilibrium structure of a normal star.

Equation of State

This physical law relates gas pressure P in a given layer to the number density of particles N (particles/cm^3) and the temperature T at that radius from the center of the star. For a perfect gas, the equation of state says

$$P = NkT$$

where k is the Boltzman constant. Pressure is simply proportional to the number density of particles and to the temperature (expressed in Kelvin). The number of particles in a given volume, however, depends upon the chemical composition of the material, because particles are not identical. An atom of helium has four times the mass of an atom of hydrogen; thus, a given amount of helium has only one-fourth the number of particles as an equal amount of hydrogen. The same total mass of these two elements in an identical volume therefore will have different pressures. If the pressure from helium is responsible for the balance against gravity in a stable star, then four times as much mass in the form of helium would be necessary at the same temperature to produce the same pressure as hydrogen.

Temperature is an additional factor. At low enough temperatures, atoms are electrically neutral. At higher temperatures, atoms ionize, the electron becoming free particles in addition to the nuclei. Each ionized hydrogen atom would be represented by two particles, the nucleus (proton) and a free electron, with a corresponding change in pressure compared to neutral hydrogen. Given the chemical composition of a gas and the state of ionization of the atoms, then the mean atomic mass μ of the particles may be calculated and the equation of state expressed as follows: the mass density ρ becomes $P = \rho kT/\mu$.

Hydrostatic Equilibrium

At every layer within a stable star, there is a balance between the inward pull of gravitation and the gas pressure. This is a **stable equilibrium,** for if gravity were greater than the gas pressure, the star would contract. On the other hand, if the gas pressure were greater, then the star would expand. In a stable configuration, the two must balance. Gas pressure in any layer thus is just equal to the weight (gravitational force) on all the matter above that given layer, in the same manner that the pressure at any depth in a pool of water equals the weight of the water above that depth, hence the term **hydrostatic equilibrium.** An immediate consequence is that gas pressure must increase inward toward the center of a star.

Thermal Equilibrium

In a steady-state (not changing) or equilibrium situation, the total energy flowing outward at a given radius (the luminosity at that radius) must just equal the total energy being generated interior to that layer. Why? Because energy always flows from a hotter to a cooler region. If energy is flowing outward faster than it is being generated, then the interior is cooling; this lowers the gas pressure, and the star will shrink. But as the star shrinks, the density will increase and the release of gravitational energy will go into heating up the material. This process causes nuclear reactions to go faster, thus generating more energy. When a balance is struck between energy generation and the outflow of energy, the star has achieved a stable structure, and no additional readjustment of the structure will occur. An immediate consequence of **thermal equilibrium** is that the inner layers of a star must be hotter than the outer layers.

Energy Generation — The CNO Cycle

In a low-mass star like the Sun, energy is generated from the conversion of hydrogen to helium in the proton-proton cycle. In higher-mass stars, to balance the greater gravitational pull requires higher pressures, hence higher central temperatures. At higher temperatures, thermonuclear reactions other than those of the proton-proton cycle become important. For stars of mass greater than about two solar masses, in fact, a different cycle of thermonuclear reactions involving the elements carbon, nitrogen, and oxygen dominates the total energy production. This is the **carbon-nitrogen-oxygen cycle** (or **CNO cycle,** for short):

$$^{12}C + {}^{1}H \rightarrow {}^{13}N + \text{energy}$$

$$^{13}N \rightarrow {}^{13}C + e^{+} + \nu$$

$$^{13}C + {}^{1}H \rightarrow {}^{14}N + \text{energy}$$

$$^{14}N + {}^{1}H \rightarrow {}^{15}O + \text{energy}$$

$$^{15}O \rightarrow {}^{15}N + e^{+} + \nu$$

$$^{15}N + {}^{1}H \rightarrow {}^{12}C + {}^{4}He + \text{energy}$$

Opacity

The structure of a star will also be affected by how easily photons can pass through the material in any given layer. If a layer absorbs photons, then it will be heated with an increase in pressure that will expand the layer. Alternatively, if a layer is transparent and allows photons to readily escape, then that layer will be cooler with a lower gas pressure that will allow gravity to compress its material. Astronomers actually use the reciprocal of transparency, or **opacity,** to measure the ability of photons to pass through material. A low opacity means high transparency; a high opacity means a low transparency. The opacity of the material in a given layer depends on the chemical composition (some elements absorb light more easily than other elements), the temperature, and the density.

Energy Transport

Each layer in a star must further be checked for the conditions that determine whether the flow of energy is by **radiation** (photons) or by **convection,** the movement of hotter mass to cooler regions and cooler material into hotter regions. If a large temperature change occurs over a small distance in a star, then more energy is present than can be moved outward by the diffusion of photons. Thus, the matter must absorb the energy. It becomes hotter and expands, becoming lighter relative to its immediate surroundings. Lighter materials are buoyant and will rise, carrying energy in the form of heat or thermal energy outwards. Ultimately, when the matter has risen to a level where the temperature gradient is more gentle, radiation of light becomes effective in lowering the energy content of the material. The matter cools and contracts. Denser material loses its buoyancy and begins to sink to lower layers of the star. Transport of energy by convection is a very efficient means by which the star can move energy from the interior towards the surface where it can be radiated away.

High-Mass Stars versus Low-Mass Stars

The amount of energy being generated each second at any point in the interior of the star is determined by how much hydrogen is being converted into helium each second per unit mass. This process is called the **nuclear reaction rate.** The reaction rate depends on the temperature, density, and chemical composition. How fast helium is produced by the carbon-nitrogen-oxygen cycle is very different from the reaction rate of the proton-proton cycle. As a result, the CNO cycle dominates the total energy production in stars more massive than twice the solar mass. For lower-mass stars, the proton-proton cycle dominates the generation of energy.

The difference in temperature dependency of these two forms of energy generation not only affects which cycle dominates the total energy generation but also has an immediate effect on the internal structure of main sequence stars. Because of its extreme dependence on temperature, the CNO cycle dumps most of the energy generated in a high-mass star into a very tiny region about the star's center. Radiation cannot move this energy away fast enough, but convection can. In the outer part of the star where the temperature gradient is more gentle, radiation is adequate to move the energy farther out to the star's visible surface layer. On the other hand, the proton-proton cycle has a reaction rate that varies relatively gently with temperatures. As a direct consequence, the energy produced in a low-mass star occurs over a large part of the interior of the star. The temperature gradient is low, and radiation is able to carry away the energy. In the outer, cooler parts of the star, however, photons are absorbed:

neutral atom + photon → ionized atom + free electron + kinetic energy

Only part of the energy of an absorbed photon goes into breaking the bond between the atomic nucleus and the electron; the rest becomes energy of motion. Faster moving atoms mean a higher temperature; the matter thus expands producing the condition for convection. In the outer layers of a low-mass star, the dominant mode of energy transport becomes convective motion. The internal structures of high-mass and low-mass stars are thus essentially reversed from each other (see Figure 10-1).

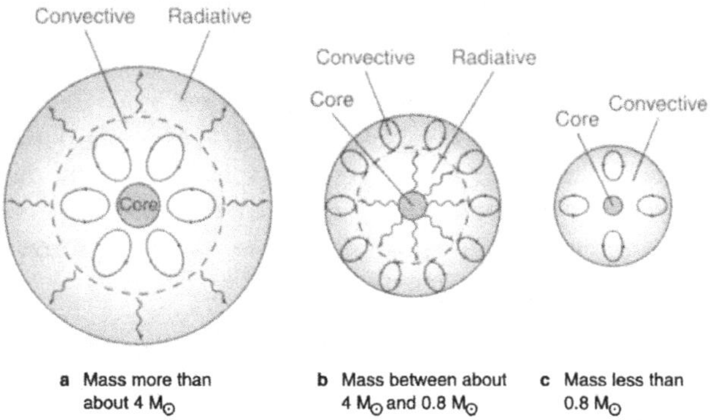

Figure 10-1

High-mass versus low-mass main sequence structure.

Other Types of Stars

In addition to ordinary stars like our Sun, the universe also contains other types of stars whose structures may differ because they exist in a multiple-star system or they produce variable energy in their cores. Some of these different types of stars include binary stars and variable stars.

Binary stars

There are many single or isolated stars like the Sun, but about half of all stars in the sky are found in multiple systems. Of the 25 nearest star systems within 4 pc (13 ly) of the Sun, 8 actually are multiple systems (7 binaries and 1 triple system). Binary systems are of special interest, because analysis of their orbital characteristics by use of Kepler's Third Law yields a direct measure of stellar masses. Such stars that are well separated are known as **visual binaries,** but others may be detected only via the Doppler Effect and hence are known as **spectroscopic binaries.** If the orientation of the orbit is such that the stars alternatively pass in front of each other, then an **eclipsing binary** is observed; analysis of the light curves yields information directly about stellar sizes.

Other phenomena are found in close binary systems. Very close stars may have their spherical structures altered by the gravitational effects of their companions. If the stars are close enough, this process may result in the near sides of the stars touching as **contact binaries.** As one or the other of the stars in a pair evolves, there may result mass exchange between the two stars that alters the course of evolution for both stars. The most dramatic examples of mass exchange are represented by the novae and x-ray binary stars. Given the large range of stellar properties, an extremely great variety of pair types and interactions is possible.

Variable stars

Stars whose luminosity changes in a periodic or non-periodic fashion are known as **variable stars.** There are dozens of different types of variables known. Among the more important are very young stars (T Tauri variables) that are in the process of establishing stable thermonuclear energy production as main sequence stars; pulsating variables whose outer layers literally swell and contract; and several types of red giant stars. The variability of any star yields clues to its internal properties (in the same fashion that differences in vibration clearly distinguish a small, lightweight snare drum from a large, heavy kettle drum), but specific types of variables are of intense interest because they can be used as distance tools.

Instability strip. A number of types of variables are known as **pulsating variables** as their outer layers swell and shrink in a regular, cyclic pattern. When distended, the pressure in the outer layers is not adequate to balance gravitation, and thus gravity will reverse their expansion.

When compressed, pressure can overbalance gravity and cause the star to re-expand. Such a pulsation is analogous to a child on a swing set; energy must be continually added to the oscillation at the proper time in each cycle to maintain an unchanging pattern of swings. Without such an addition, the ordered energy of the pulsational cycle would die out as the energy is dissipated by frictional forces into random heat.

In a star the only energy that may be tapped to add into a pulsational cycle is the flow of energy outward. The ability to tap such energy depends upon how much energy is flowing and where in the outer envelope there exists a means of using that energy. If the means exists, but is too far out in the star, there is no star left to oscillate; if too deep in the star, then there is too much overlying star to affect. At temperatures and luminosities within a band that cuts diagonally upwards across the HR diagram (see Figure 10-2), the **instability strip,** all the necessary factors are present to produce a stable cycle of oscillation. The energy-tapping mechanism is the ionization of helium that already has lost one electron:

$He^+ + \text{photon} \rightarrow He^{++} + e^- + \text{kinetic energy}$

Only for stars within the instability strip does this occur at the right time in the cycle. If a star like the Sun were to be disturbed (say, by distending it so that pressure no longer balanced gravitation), no stable oscillation would be produced because the energy of the disturbance would rapidly be converted into random motions within the stellar material.

Classical Cepheid variables. High-mass stars, once they have exhausted their core hydrogen, evolve to the right in the HR diagram. When these stars have luminosities and surface temperatures that place them within the instability strip, they will develop pulsations that affect not only their size but their surface temperatures and luminosities. The **light curves** will have a characteristic form showing a steep increase in brightness followed by a slower decrease in brightness. Any variable with this form of light variation is termed a **Cepheid variable,** after the first star of this class, δ Cephei. More specifically, a young, massive star with solar metal abundance that has recently left the main sequence and moved into the yellow supergiant region of the HR diagram is termed a **Classical** or **Type I Cepheid.** The pole star, Polaris, is an example of this type of variable star.

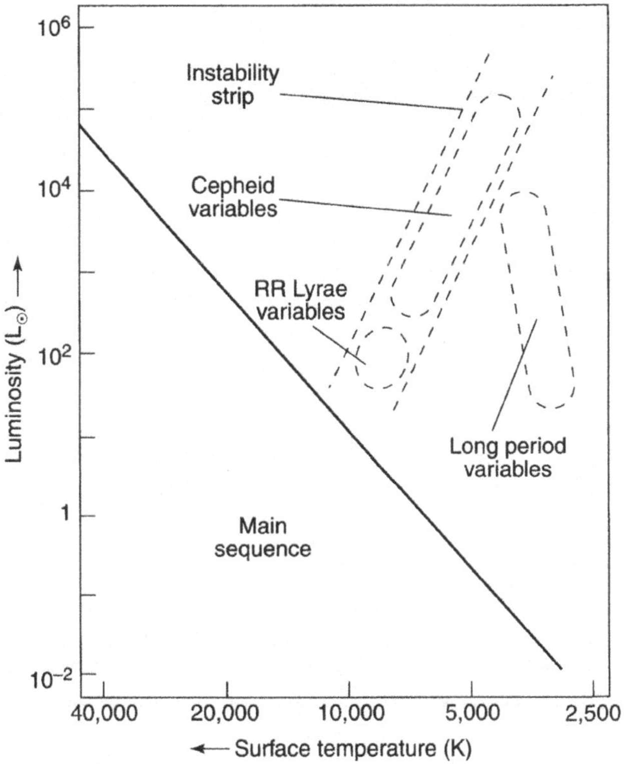

Figure 10-2

Variable stars in the HR diagram—instability strip, miras, T Tauri variables.

These Cepheids typically have periods of variability from a few days to as long as 150 days. Their luminosities are high, with absolute magnitudes between –1 to –7, and a difference between maximum and minimum light, of amplitude, of up to 1.2 magnitudes (a factor of 4 in luminosity). A Cepheid is brightest when it is expanding most rapidly, and faintest when contracting the fastest.

W Virginis variables. Young massive stars are not the only stars that can move into the region of the instability strip during some stage of their evolution. A very old, low-mass star that is between its horizontal branch stage and its planetary nebulae stage can achieve the right luminosity and surface temperature when its helium-burning shell has collided from below with its hydrogen-burning shell, temporarily ending both types of thermonuclear reactions. When this phenomenon occurs, the star undergoes a quick contraction with a rise in surface temperature that takes it leftwards across the HR diagram into the region of the instability strip. Such a star is a **Type II Cepheid** or **W Virginis star.** Typically, the periods of variability of W Virginis stars are between 12 and 20 days. Although such a star may have a luminosity and surface temperature identical to a Classical Cepheid, their periods will be different.

RR Lyrae variables. The third major class of variable with a Cepheid-like light curve is the **RR Lyrae variables** (also called cluster variables, because they are common in the globular star clusters). These stars have short periods, between 1.5 hours to 24 hours. They are fainter than the Cepheids, with luminosities of about 40 times that of the Sun. Like the W Virginis stars, these are old, low-mass stars, specifically horizontal branch stars (core helium-burning stars) whose surface temperature places them within the bounds of the instability strip.

Period Luminosity Relationship. A fundamental importance of the Cepheids is the existence of a relationship between their period of pulsation and their intrinsic luminosity, originally discovered by Henrietta Leavitt from a study of these variable stars in the Large and Small Magellanic Clouds. The **period luminosity relationship** differs for the Classical Cepheids and the W Virginis stars, with the former being about four times more luminous at any given period. Determination of the period of variability for any star is fairly straightforward, and once that period is known, the intrinsic luminosity of the star may be deduced. Comparison with the apparent brightness of the star then yields the distance to the star. As these are intrinsically very bright stars, they can be identified at distances as great as 20,000,000 parsecs, making them an extremely valuable tool for obtaining distances to a large sample of nearby galaxies. Indeed, they are a critical key to getting the distance scale of the Universe.

Irregular, semi-regular, and Mira variables. A second important class of variables is the red variables. These stars do not have a stable cycle of variability, but exhibit semi-regular or irregular behavior with periods of a few months to about two years, again due to deep ionization regions. In the highly distended outer parts of these stars, energy absorbed and released by ionization can produce shock waves that dramatically affect the surface layers, producing strong stellar winds with mass loss up to 10^{-5} solar masses per year. In addition, condensation of molecules into dust grains can further obscure the light coming from these stars.

A prime example is the star Mira (the name means "wondress") whose visible light varies by a factor of 100 in a semi-regular manner over an approximate 330-day period. Its total luminosity variation is only a factor of 2, but the greater part of that radiation is in the invisible infrared part of the spectrum. The variation of temperature over its cycle, with the peak wavelength of its radiation in the infrared, results in a major change in visible brightness.

CHAPTER 11
FORMATION AND EVOLUTION OF STARS

Formation of Stars

Interstellar space is filled with diffuse gas and dust. Relatively denser and cooler regions, up to 50 pc in diameter and with a million solar masses, are filled with molecules. In these **molecular clouds,** shock fronts from nearby star formations or a supernova explosion or some other global gravitational disturbance may begin the process of self-gravitational contraction, leading to the formation of new stars. The earliest stages of pre–main sequence evolution are not directly observable, because **protostars** are hidden behind massive amounts of dust. Consequently, no radiation from the forming star is visible. If one could envision a protostar without the obscuring dust, theory suggests that initially a protostar would be very cool but luminous, with convection very efficiently moving gravitational energy that is released in the interior outwards to the exterior. As the object shrank, the surface area would dramatically decrease and the overall luminosity likewise decrease rapidly.

Bok globules, infrared stars, and cocoon stars

Some of these earliest stages of evolution are believed to occur in the small, dense, dark dust clouds that often are seen silhouetted against more extended regions of luminous, hot, interstellar nebulae. These are the **Bok globules.** Observation of radio radiation that penetrates the dust from these sites suggests that internal motions of the interstellar material are in a stage of contraction. Such an object may also be termed a **cocoon star,** because of the surrounding shroud of dense, opaque dust. When the dust is sufficiently warmed by radiation from the interior protostar, it in turn will radiate in the infrared. Many **infrared sources** are observed in regions where star formation is taking place. This stage of evolution is also termed **Helmholtz contraction**—one-half of the energy released by gravitation contraction into the protostellar material results in heating, and one-half of the energy is convected to the surface to be radiated away.

As the core temperature of the protostar rises, ionization of the material occurs. Photons are not absorbed as well by ionized material, thus a transparent **radiative core** forms in which the energy is transported by photons. Photons, however, cannot directly move to the surface because they are continually colliding with the nuclei and electrons. In a collision, a photon's direction is changed; it is just as likely to be reflected back into the interior as not. Photons slowly drift outwards to the surface, but along the way each undergoes a tremendous number of collisions, and the time to ultimate escape is large, amounting to as long as a million years or more. At this stage the luminosity reaching the surface has declined greatly, but now starts to slowly rise as the protostar continues its contraction. The protostar's surface temperature also increases, thus it now moves parallel to the main sequence in the HR diagram (see Figure 11-1). When the central temperature rises to about 10^6 K, the first energy generation via nuclear reactions commences. By this time the outer layers of surrounding material may be blown away, revealing a new star.

T Tauri variable stars

A star does not become stable instantaneously, however. By analogy, imagine a rubber ball sitting on a table. The ball is in a state of equilibrium, with the surface tension of the tabletop pushing up on the ball to balance the downward gravitation force exerted by Earth. When dropped from a height, the ball bounces around for a while before that final equilibrium is reached. A new star is similar: Gravitation contracts the star, heating the central material until thermonuclear reactions begin. But at this point, the star is overcontracted. The new source of energy produces sufficient heating to increase central pressure and overbalance gravity. Thus the central core begins to expand, but expansion is accompanied by a cooling that damps the thermonuclear reactions. Now pressure is too low to balance gravity, and core contraction begins anew. This "bouncing" of the central core around its ultimate state of equilibrium is matched by changes in the gravitation-pressure balance in the outer part of the star. The surface responds to variations in the core with erratic variations in the surface radius, temperature, and luminosity, producing a **T Tauri variable** star. FU Orionis is an example of such a young variable star. FU Orionis was discovered in 1939 only when its surrounding dust became sufficiently transparent for its light to be seen.

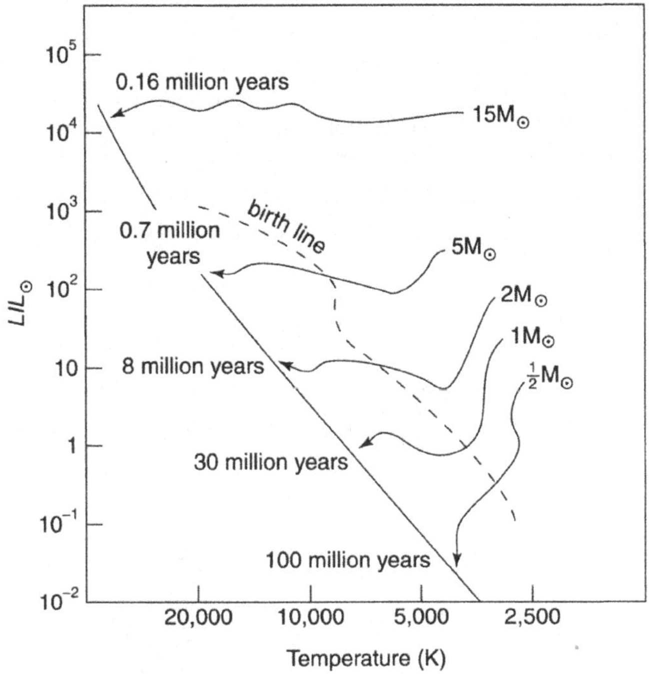

Figure 11-1

HR diagram showing pre–main sequence evolution for stars of different masses. The dashed line separates those stages (at right) that are hidden within dense clouds of dust and thus not directly observable from those stages (at left) that can be directly seen when the surrounding dust has been blown away and dispersed.

In general properties, T Tauri stars are young, typically 10^5 to 10^6 years old, and usually found in widely dispersed groups (**T associations**) embedded in regions of gas and dust (for example, in the direction of the Orion Nebula). Their brightnesses and surface temperatures place them to the right of the main sequence (see Figure 11-2).

They are erratically variable, with amplitudes up to 3 magnitudes (a factor of 16 between minimum and maximum brightness). Their spectra show intense emission lines from extensive outer coronas indicating substantial mass loss in strong stellar winds at velocities of 70 to 200 km/s. Star formation is not a very efficient process in converting interstellar material into new stars.

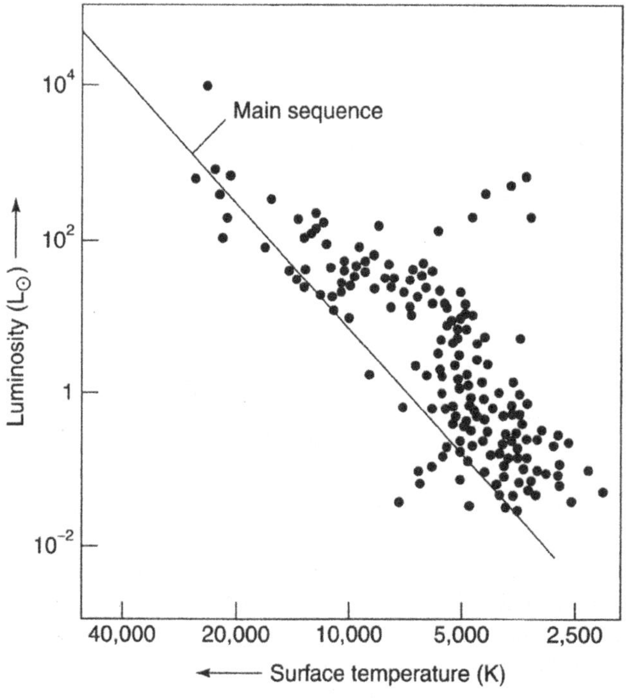

Figure 11-2

T Tauri stars in the HR diagram.

Only relatively low mass stars are observed in their T Tauri stage. Higher-mass stars with their greater gravity pass through this stage so quickly that the likelihood of actually observing a high-mass T Tauri star is very small. A protostar on its way to becoming a 17 solar mass B0 star, for example, may take as little as 100,000 years to contract and achieve main sequence stability. A star like the Sun contracts much more slowly, taking about 30,000,000 years to become a full fledged main sequence star, whereas a low-mass M5 star (0.2 solar masses) could take 500,000,000 years to accomplish its main sequence stage of stability.

Related to T Tauri stars are the **Herbig-Haro objects.** Originally discovered as compact regions of emission from hot interstellar gas, these were soon found to occur in pairs, on opposite sides of an often unseen star in its stage of becoming a main sequence star. Mass loss often occurs in the form of a **bipolar outflow** from protostars, two jets in opposite directions. When these jets collide with denser regions in the surrounding interstellar material, the gas is heated and can be observed. Herbig-Haro objects were initially thought to be very young stars, but observations made using the Hubble Space Telescope in 1999 show bipolar outflows in stars that appear to be forming planets, which usually happens when the star is much older. The star HD163296 is one such object; it shows bipolar jets as well as evidence of planetary formation.

Evolution of Stars

The interior of a typical main sequence star is illustrated by the internal conditions of the Sun, with the highest density, pressure, energy generation rate, and temperature occurring at the very center. The temperature dependency of the proton-proton cycle means that energy is produced over a fairly large volume in the stellar center, out to about 25 percent of the total stellar radius in a star like the Sun.

Within this core, the star's chemical composition slowly changes as hydrogen is converted to helium (see Figure 11-3). In the 4.6 billion

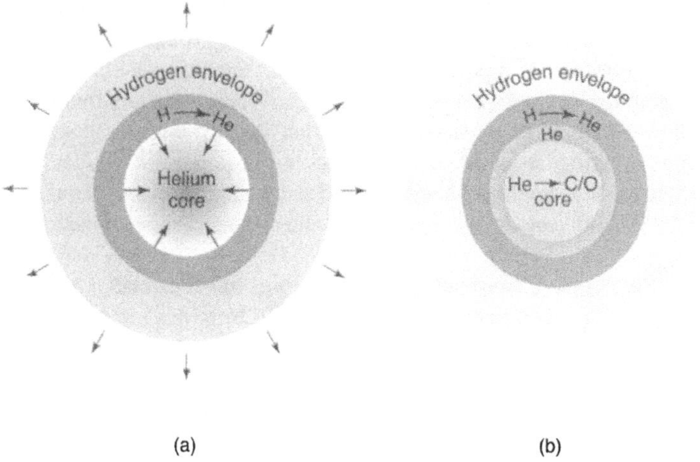

Figure 11-3

Representative stages in post–Main Sequence evolution. At left, the star's core has been converted to helium and is slowly shrinking. Exterior to the core is a shell-like region in which hydrogen is converting to helium. The envelope of the star is still dominated by hydrogen and slowly expands as this star evolves into the red giant region. At right, central temperatures have reached the point where helium can convert into carbon and oxygen in the core. This is surrounded by a shell that is not hot enough for helium reactions. Surrounding this is a shell in which hydrogen is being converted to helium. The outermost layer is the hydrogen-dominated envelope. (Sizes of regions are not drawn to scale.)

years since the Sun formed, it has used about one-half of its hydrogen at the very center. This slow progressive conversion of lightweight hydrogen to fewer nuclei of heavier hydrogen is accompanied by slow changes in other physical factors in the stellar interior and related slow changes in the star's surface conditions. In due course, when all hydrogen in the core is exhausted, a star must make more dramatic changes in its structure. To a biologist, changes that occur in the lifetime of a living organism are referred to as aging. Astronomers refer to the aging of a star as **stellar evolution.**

Zero-Age Main Sequence

Upon the onset of central thermonuclear reactions, a star's chemical composition is homogeneous throughout its interior. The equilibrium structure of such a star is one whose surface temperature is a bit warmer and whose luminosity is a bit smaller than a typical star in that portion of the main sequence. The locus of newly formed, chemically homogeneous **zero-age main sequence** (**ZAMS**) stars forms a boundary to the lower left of the main sequence. As a star ages, lightweight hydrogen is converted to fewer nuclei of heavier helium. To maintain sufficient central pressure to balance gravity, both the mass density must increase slowly as well as the central temperature. Mass density can increase only if the stellar core slowly contracts (such an equilibrium that slowly changes is called **quasi-static equilibrium**). Although the fraction of hydrogen is decreasing, the higher temperature ensures a somewhat greater rate of energy production that means a higher surface luminosity. The increase in interior temperature is accompanied by an increase in temperature in layers outside of the core of the star. Here, in the deeper **stellar envelope,** the higher temperature produces a higher pressure that expands the outer part of the star. Near the surface, however, the expansion supported from below is accompanied by a decrease in temperature. In sum, as the hydrogen is consumed in the central core, the stellar photosphere cools slightly and the luminosity increases (in other words, the star drifts rightward and upward across the main sequence band in the HR diagram).

Terminal-Age Main Sequence

Stars that are at the point of exhausting hydrogen in their cores form a locus that bounds the main sequence band on the upper right. As these stars are at the end of their main sequence stage, they are termed **terminal-age main sequence** (TAMS) stars. (See Figure 11-4.)

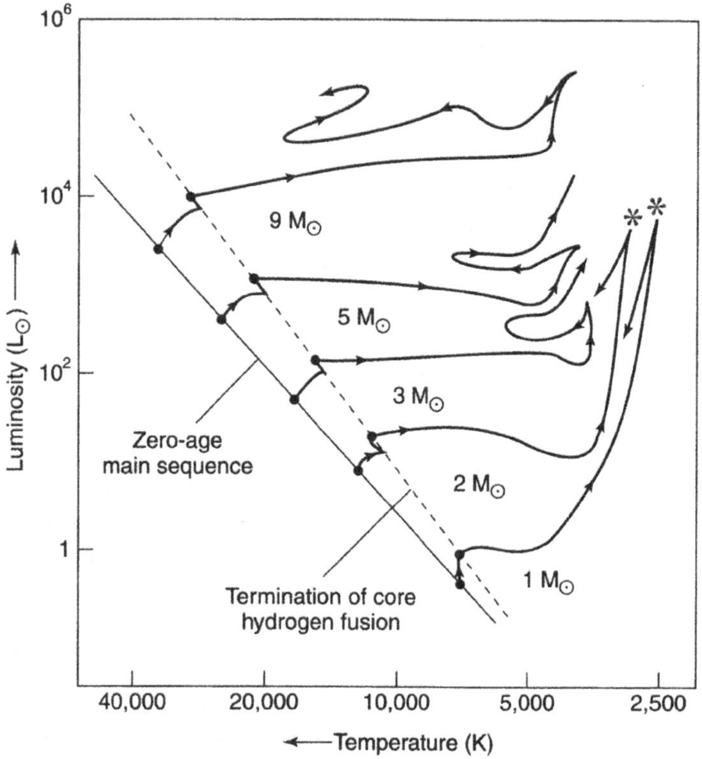

Figure 11-4

Evolution across the main sequence and into the giant region; the zero-age main sequence and terminal-age main sequence configurations are marked. The asterisks mark the stages for low-mass stars at which time helium explosively begins to convert to carbon in the stellar core, resulting in a quick change in the stars' luminosities and surface temperatures.

Main Sequence Lifetimes

The luminosity of a star is a measure of how fast a star is using its nuclear fuel. The mass of a star indicates how much fuel is available. The lifetime of a star in any given evolutionary stage is given by the amount of available fuel for that stage divided by the rate of consumption of that fuel; in other words, lifetime is proportional to the mass divided by the luminosity. Because the mass-luminosity relation for main sequence stars shows that luminosity is proportional to $mass^{3.5}$, a star's lifetime is proportional to $mass^{-2.5}$. Bright, massive main sequence stars must evolve faster than faint, low-mass stars. Not only are these stars intrinsically rarer than lower-mass stars, but they do not last as long. More quantitative main sequence lifetimes may be obtained from theoretical calculations as shown in Table 11-1.

Table 11-1: Approximate Main Sequence Lifetimes

Stellar Mass	*Main Sequence Lifetime*
50 $M_\odot$	5×10^5 years
25 $M_\odot$	3×10^6 years
10 $M_\odot$	3×10^7 years
2 $M_\odot$	2×10^9 years
1 $M_\odot$	9×10^9 years
0.5 $M_\odot$	6×10^{10} years
0.1 $M_\odot$	3×10^{12} years

Red Giant Stage

What happens when the core hydrogen is depleted in a main sequence star like the Sun? In such a star, energy is still flowing outward from the core. The only means of replacing this lost energy is through gravitational contraction of the core, with half the released energy going into heat, the other half moving outward in the star to be radiated away at the surface. Density, temperature, and pressure all increase at the center; but the outer envelope responds to changes in the core by expanding. The surface area increases as the surface temperature decreases, yielding a constant luminosity. In the HR diagram, the star appears on observation to be "moving" to the right of the main sequence.

Immediately outside the helium core, the stellar material is still hydrogen-rich. As the layer just above the helium contracts, it becomes hotter and denser until a **shell-burning source** in which thermonuclear reactions converting hydrogen to helium are established. The star can now be considered to have three distinct regions—a central spherical core that is made of helium, a thin layer above the core in which hydrogen is being converted to helium, and lastly the thick outer envelope of the star comprised of the proportion of hydrogen (about 74 percent) and helium (about 24 percent) with which the star formed. The core contraction continues, the hydrogen-burning shell slowly moves outward, and the exterior of the star continues to expand and cool, but the stellar luminosity remains approximately constant. The star is now a **sub-giant star,** on its way to becoming a red giant.

The whole outer envelope of the star experiences decreasing temperatures. Of specific interest to astronomers is the depth of the interior layer, which sets the transition from a radiative interior to the convective envelope. When the surface temperature drops to around 3,000 K, this transition layer begins to move into the interior of the star; thus the depth of the outer convective envelope increases. Convection, however, is a very efficient means of moving energy outward, whereas the movement of energy by the diffusion of photons is slow. The interior radiative zone seems to store the energy produced by the core of the star over the previous few million years, because it is trapped by the slow radiative process. As the convection moves into the interior, this stored energy is brought rapidly to the surface and the star begins to brighten rapidly, though the surface remains at a constant temperature. Observationally, the star is seen moving up the **Hayashi track,** named after the Japanese theoretician who first recognized this stage of stellar evolution. This stage is also called the **red giant branch.** The core in turn responds to this outer loss of energy by accelerating its contraction. At the same time, the shell burning accelerates and the size of the helium core grows in mass more quickly.

Triple-alpha process

As long as the core temperature is below about 200 million K, the only source of energy to replace that which is flowing outward into the cooler envelope is gravitational contraction in the center and hydrogen fusion in a shell just outside the helium core. Thermonuclear reactions uniting two helium nuclei into one of beryllium do occur, but beryllium is unstable; as soon as a beryllium nucleus is produced, it immediately splits to give back two helium nuclei. As the central temperatures and density continue to rise, the reaction $He^4 + He^4 \rightarrow Be^8$ proceeds at an ever fast rate, but so does the reverse reaction $Be^8 \rightarrow He^4 + He^4$. But as these reactions occur faster, the ambient amount of short-lived beryllium is growing larger. When the central temperature approaches 200,000,000 K, there is sufficient beryllium at any instance that the likelihood of a reaction with another helium nucleus becomes significant.

$$He^4 + Be^8 \rightarrow C^{12} + \text{energy}$$

Because the production of stable carbon requires essentially a simultaneous collision of three helium nuclei (also known as **alpha particles**), this is termed the **triple-alpha process,** also called **helium burning.** For a Sun-like star, it has been about a billion years since the star left the main sequence to bring the core to a circumstance where a new source of thermonuclear energy can be established. But a new factor has also come into play as the helium core became denser and denser: When pushed together until they essentially touch each other, the electrons resist any further compression. Their resistance is an **electron degeneracy pressure** that adds to and actually becomes greater than normal gas pressure. Unlike normal gas pressure, degeneracy pressure does not depend on temperature. Because the helium core is degenerate, the onset of helium-burning is not moderated. Under normal circumstances, energy from nuclear fusion would heat the material, the pressure would rise forcing an expansion, which then would damp down the reaction rate. But in degenerate material, the onset of helium-burning is so rapid as to be virtually explosive, a **helium flash.** But the energy produced in the flash quickly expands the core region and removes the degeneracy (that is, normal gas pressure becomes dominant), allowing a stable fusion of helium to carbon to ensue. The whole structure of the star readjusts dramatically to this new condition. With thermonuclear energy production replacing gravitational energy in the core, the exterior of the star shrinks and gains in temperature. It becomes a smaller and hotter type of red giant star.

Stellar mass loss

A second factor, mass loss, is important during the star's evolution on the Hayashi track. A one solar mass star likely loses between 10 percent and 60 percent of its mass through strong stellar winds. The explosive onset of helium burning may also produce shock waves that blow off additional mass at the surface. The properties of the resultant star depend greatly on how much mass the star has retained.

Horizontal branch stars

In its core helium-burning phase, a 0.9 solar mass star has a luminosity about 40 times that of the present-day Sun and a relatively cool surface temperature. A 0.4 solar mass star is smaller and hotter, but still has about the same luminosity. Because luminosity does not depend on the resultant mass, but surface temperature and radius do, these core helium-burning stars are called **horizontal branch stars**—plotted in an HR diagram, they fall in a horizontal track across the diagram (see Figure 11-5).

Old globular clusters in the galaxy show low-mass stars still in their main sequence stages, whereas slightly more massive stars are found at various stages along the evolutionary part to the red giant and horizontal branch states of evolution. This turnoff of the main sequence can be used to determine the age of the cluster, because the mass of the star at the turnoff determines how long it has been on the main sequence.

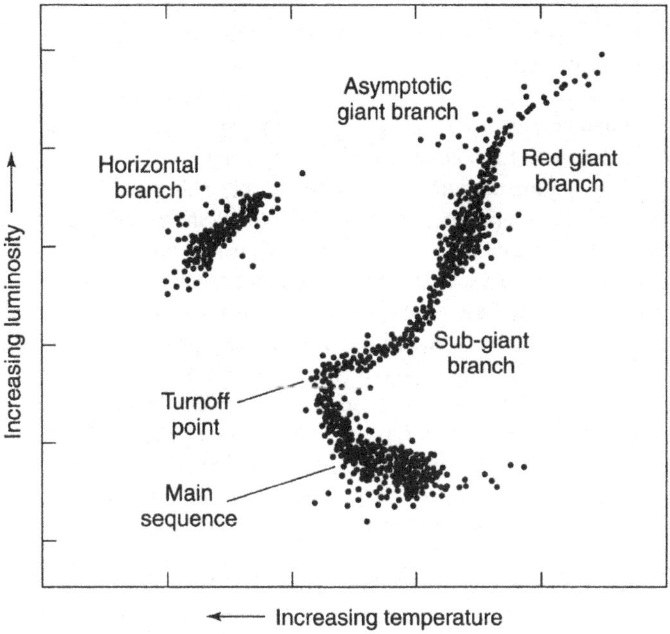

Figure 11-5

Schematic globular cluster HR diagram with branches labeled.

All nuclear reactions do not produce the same energy. The triple-alpha process 3 $He^4 \rightarrow C^{12}$ generates a relatively small amount of energy compared to hydrogen burning 4 $H^1 \rightarrow He^4$. As the carbon content builds up in the core, oxygen is produced via

$$He^4 + C^{12} \rightarrow O^{16} + \text{energy}$$

But this production contributes only a small amount of additional energy. As this stage of evolution is a reasonably bright giant star, the nucleus fuel must also be used more rapidly by the star. Hence the horizontal branch lifetime is relatively short, about 100 million years (or about 1 percent of the hydrogen-burning main sequence lifetime for the same star).

When helium becomes depleted in the core, the star again reverts to gravitation as the source of energy to replace that flowing out of the core. Once again, the surface swells to become a larger, cooler red giant star that follows a luminosity-temperature track in the HR-diagram just above the red giant branch. (As their luminosity and temperature pass very close to the track of the pre-horizontal branch red giants, they are known as **asymptotic branch stars.**) Immediately outside the inert carbon-oxygen core, a shell in which helium converts to carbon and oxygen is established. Farther out in the star there is a shell in which helium is at too low a temperature to support thermonuclear reactions; above that, hydrogen reactions produce helium in another layer. These shells in a sense resemble the layers in the interior of an onion, and computer models designed to reproduce the conditions in real stars are referred to as **onion-shell models.**

In these low-mass stars, there is insufficient gravity to compress the core to even higher temperatures and densities that would allow thermonuclear reactions producing even heavy elements. Once again at some point the existence of the electrons becomes increasingly important. As the carbon-oxygen core grows in size and continues its slow contraction, the electrons again begin to resist further compression and electron degeneracy pressure dominates the balance against gravity. The electrons eventually halt any further contraction of the star, leading to its final state.

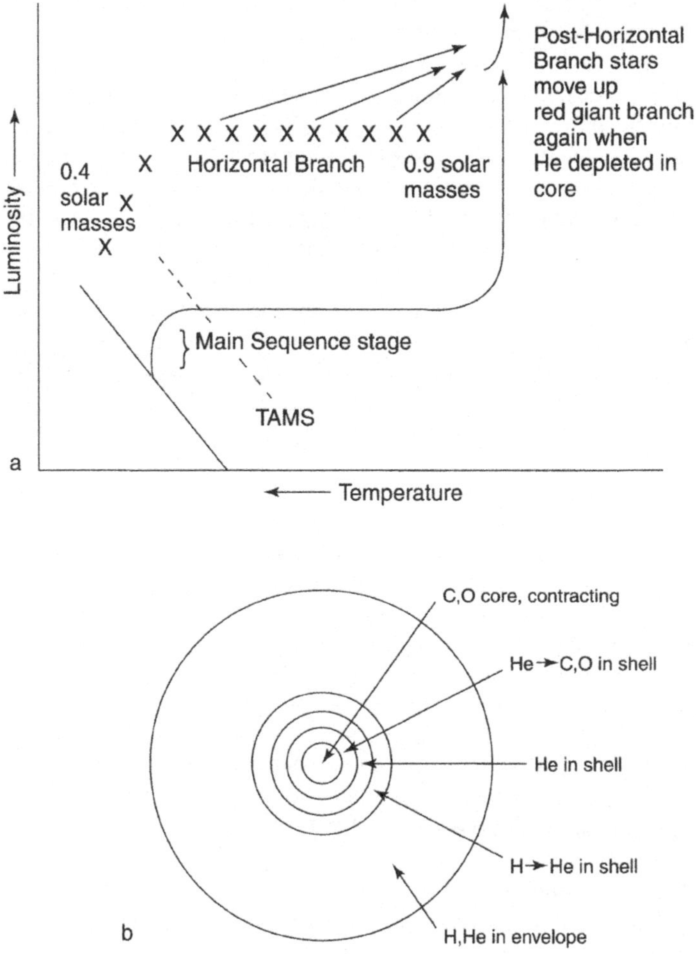

Figure 11-6a and 11-6b

a) HR diagram of post-horizontal branch evolution; b) Cross-sectional model of post-horizontal branch star configuration. (Not drawn to scale.)

The post-horizontal branch evolution of a low-mass star is complicated. The greater part of the star's luminosity is coming out of the slowly shrinking core, with additional contributions from the helium-burning and hydrogen-burning shells that are slowly moving their way outward into the material of the stellar envelope. Outer convection rapidly takes this energy to the surface, and the star moves up the Hayashi track to become an extremely luminous red supergiant star. The outward movement of the hydrogen-burning shell is slower than that of the helium-burning shell because helium-burning produces so much less energy per unit mass — helium must be processed more quickly to supply the energy to maintain the inner stability of the star. But at some point, the helium-burning shell must overtake from below the hydrogen-burning shell. When this happens, the hydrogen thermonuclear reactions shut down, but then so do the helium reactions because there now is insufficient helium to support them.

This loss of energy to the outward flow of luminosity produces an immediate effect in the outer regions and at the surface of the star. The surface quickly shrinks and becomes hotter — in terms of where the star would be plotted in the HR diagram, it undergoes a very rapid excursion into the blue supergiant region. Readjustment of the outer part of the star allows the hydrogen-burning shell to be reestablished and to produce a new shell of helium below it. The outer layers readjust, with the surface expanding and cooling once again to take the star back into the red supergiant region. Such an excursion across the HR diagram is very rapid, lasting no more than a few thousand years. As a consequence, stars undergoing this type of shell-burning instability are rarely observed.

Planetary nebulae

Within a few million years, the greater part of the stellar mass is now converted to carbon and oxygen. The hydrogen- and helium-burning shells have now moved so far out into the exterior of the star that little material remains above these layers. In fact, at some point these energy-producing regions heat the outer stellar material so much that the star's gravity is no longer able to retain the outermost layers. These layers now are simply shed, blowing outward in a stellar wind, sometimes forming a spherically symmetric nebulosity about the remaining star (the Ring Nebula is the most familiar example) and other times forming more complex patterns of nebulosity. At the centers of these **planetary nebulae** (so named because when originally discovered, astronomers erroneously assumed these nebulae were planetary systems in formation, not dying stars as they actually are) are located the very hot interior remnant of the original star. This, now bereft of thermonuclear reactions, rapidly shrinks and becomes much dimmer as it finally stabilizes into a white dwarf.

Large numbers of planetary nebulae are known, and each contains a very hot central star surrounded by a complex nebula. Initially, when the star is a red giant, a slow and dense wind blows outward, and is usually more or less spherical (though it can be significantly flattened). However, as the star loses mass, the hotter core is exposed, and a faster, less dense wind is blown. This catches up to and collides with the slower denser wind, shaping it into a variety of shapes, depending on the geometery of the winds and our viewing angle. These planetary nebulae can be shaped like barrels, spheres, or even butterfly shapes with two huge lobes looking like the wings of a butterfly. The famous Ring Nebula is actually a barrel-shaped structure, and we happen to be peering almost directly down the long axis, making the nebula appear round. Typical diameters of the nebula are 20,000 to 100,000 AU in size, representing 0.1 to 0.2 solar masses of material, with expansion velocities usually 10 to 12 km/s, although higher velocities have been observed. Likely all stars less than a few solar masses in their main sequence stars ultimately go through the planetary nebula stage, by which time the star retains no more than about a solar mass of material. Such planetary nebulae stars are observed with temperatures of 50,000 to 200,000 K (measured for the object NGC 2440), but their sizes are much smaller than the Sun. The larger of these stars are associated with the smaller nebulae, and the smaller stars are found in the larger nebulae, clearly indicating that these stars are shrinking as the surrounding nebulae expand over time. The observable lifetime of this phenomenon may be only 20,000 years as these stars undergo their final restructuring into a white dwarf configuration. See Figure 11-7.

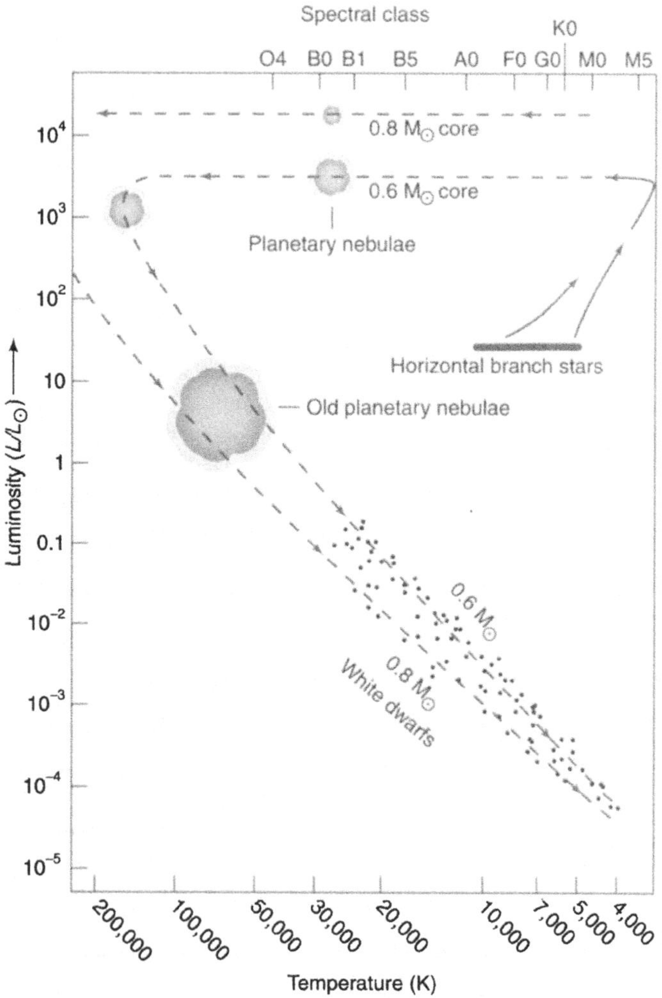

Figure 11-7

HR diagram of the planetary nebulae stage and white dwarfs.

CHAPTER 12
FINAL END STATES OF STARS

Every star visible in the night sky is in some active phase of its existence, converting lighter elements into heavier ones as their primary source of energy. Ultimately, this process must come to an end, and each star will settle into some final state. Low-mass stars peaceably become compact white dwarf stars. Higher-mass stars, however, explosively end their active phases, leaving behind even more compact objects, neutron stars, and black holes.

White Dwarf Stars

The first **white dwarf star** to be discovered was the companion (component B) of Sirius, a main sequence star (component A). The spectrum and color, hence temperature, of the two stars are very similar, but the dwarf star is some 10,000 times fainter, hence much smaller by factor of 100 in radius, than the main sequence companion. With a mass of about one solar mass and a size about that of Earth, this star has an average mass density of 3,000,000 grams per cubic centimeter. Its surface gravity is such that a person who weighs 180 pounds on Earth would weight 50,000 tons on the surface of Sirius B.

Numerous white dwarfs are known (their number is second only to the main sequence stars) and as a class, they are hot stars with typical surface temperatures of 10,000 K but luminosities less than about 1 percent of the Sun. These stars have no internal thermonuclear energy source. The energy they radiate is the loss of their residual heat. This loss, however, is slow due to their small surface areas. Comparing luminosities with their (theoretical) store of internal thermal energy suggests that cooling times are long—from 10 to 100 billion years. Ultimately they will end up as black dwarfs, too cold to have any significant radiation.

White dwarfs are the final stage of the evolution of **low-mass stars,** which (according to current theory) originally had no more than about 8 solar masses while on the main sequence. Mass loss during the red giant and planetary nebula stages can reduce this to no more than about 1.4 solar masses. This number is known as **the Chandrasekhar limit,** or the maximum mass for which the electron degeneracy pressure is able to balance the inward pull of gravity in a star of this size. More massive main sequence stars tend to produce more massive white dwarfs, with the most common mass being about 0.6 solar mass. This remaining mass has been converted basically to carbon and oxygen. Unlike a normal star in which gas pressure from the motion of all the particles—nuclei and electrons—provides the balance against gravity, in a white dwarf the nuclei of carbon and oxygen play little role. Ultimately, when they cool below 4,000 K, these nuclei will lock into a solid lattice structure, the electrons providing the pressure to balance gravity. There is a limit to which this works. Unlike main sequence stars that are larger for greater mass, increasing the mass of a white dwarf star produces a smaller star. If the mass is made larger, the star's larger gravity produces a smaller white dwarf. At a mass greater than the Chandrasekhar limit, the electrons are unable to provide sufficient pressure to balance gravity. Gravity would thus compact the object to a much smaller size; but this would no longer be a white dwarf. As the properties of white dwarfs are so different from those of normal stars they sometimes are referred to not as stars but as **compact objects.** See Figure 12-1.

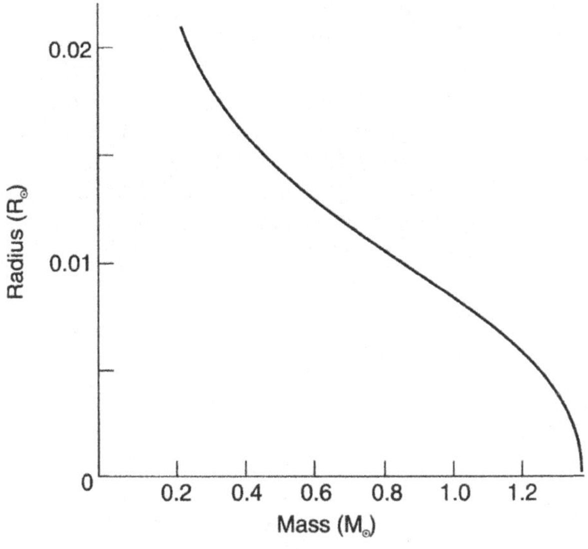

Figure 12-1

Mass-radius relation for white dwarf stars.

Novae

A white dwarf in a binary system may find itself in a circumstance where it gains mass from its companion star. When that companion evolves to become a red giant, it may become large enough that portions of its outer atmosphere are attracted by the gravity of the orbiting white dwarf, which begins to capture this hydrogen rich material onto its surface. This is not a stable circumstance. When a critical point is reached in the accumulation of this new matter, thermonuclear reactions at the surface can be set off that explosively convert the hydrogen to helium. Observationally, such an event, a **nova,** would be seen as a rapid rise (less than a day) in brightness of the star to 10,000–100,000 solar luminosities, followed by a decline in brightness over the next few months. While some mass (usually much less than 0.01 solar mass) might be shed at high velocity up to 2,000 km/s, neither the basic structure of the white dwarf nor its orbit relative to its companion are affected in this event. Mass accretion would resume until perhaps 50 to 100 years later; the accumulated material again would undergo rapid thermonuclear reactions in another nova. A number of such stars have been observed to repeat and are termed **recurrent novae.**

Type I Supernovae

If mass accretion continues, interspersed with novae events, the total mass of the white dwarf may approach the Chandrasekhar limit. At that point, electron pressure can no longer balance gravity and the star begins a catastrophic collapse. Its carbon-oxygen material is compressed, and runaway thermonuclear reactions are triggered: $C^{12} + C^{12} \rightarrow Mg^{24}$ + energy; $O^{16} + O^{16} \rightarrow S^{32}$ + energy; $C^{12} + O^{16} \rightarrow Si^{28}$ + energy, and so forth, leading to the elements iron and nickel. These reactions dump so much energy into the star that *it literally is blown apart* in an explosion or **Type I supernova.** The stellar material, now enriched in large quantities of heavy elements, is returned completely to the interstellar material. In other galaxies, Type I supernovae are observed in regions dominated by old stars, which is consistent with the fact that Type I supernovae come from old, evolved white dwarfs.

Type II Supernovae

Main sequence stars more massive than 8 solar masses are unable to lose sufficient mass to become white dwarfs. Their higher luminosities drive all evolutionary stages faster. The core is quickly converted to helium and then helium reactions produce a carbon-oxygen core more massive than 1.4 solar masses, the limit at which electron pressure can balance gravity (in other words, the central temperature and density increase until ordinary gas pressure provides the balance against gravity). As electron pressure plays no role in stopping further core contraction, subsequent stages of core contraction can proceed, each momentarily halted by establishment, first in the core and then in outwardly moving spherical shells, of various thermonuclear reactions converting lighter elements into heavier elements (see Figure 12-2). Such a star will quickly evolve to an **onion-shell configuration** marked by conversion of sulfur, silicon, and magnesium in the core to iron and similar elements (see Figure 12-3). Exterior to this core is a layer of these elements at cooler temperatures and unable to convert to iron. In a higher layer, carbon and oxygen are reacting to produce sulfur, silicon, and magnesium. This is overlain by an inert region that in turn is overlain by a shell of helium converting to carbon and oxygen. Further out is a hydrogen-burning shell moving outwards into the envelope of the star.

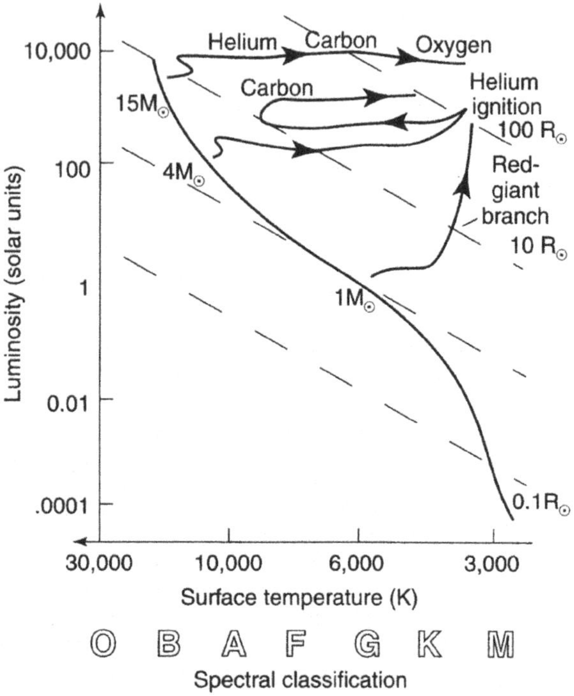

Figure 12-2

Evolutionary tracks of high-mass stars.

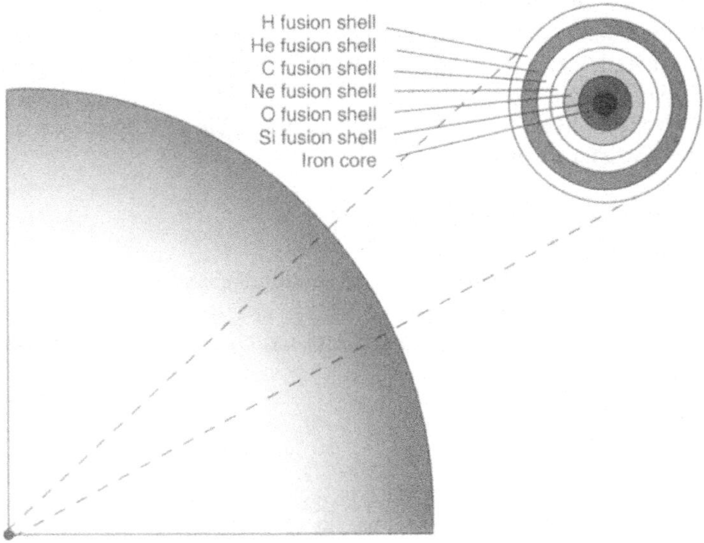

Figure 12-3

Onion shell model. Nuclear reactions are limited to the densest and hottest central region of the star.

The conversion of a stellar core to iron is a serious problem for a star, because iron is a minimum energy nuclear configuration—nuclear reactions involving iron require the input of energy. During the evolution of a star, energy was released (and radiated away) in converting lightweight elements to iron, thus to convert iron back to lightweight elements requires recapture of this energy. Similarly, the conversion of iron into heavier elements also requires energy. The star now has no thermonuclear energy source in the core, but the core is losing energy outwards because energy always flows from high-temperature to low-temperature regions. The only source of energy is gravitational, and now the core begins to contract in free-fall collapse. Within a tenth of a second, the central temperature reaches 200 billion K and the density 10^{12} g/cm^3.

The catastrophic collapse is the result of two factors. Gravitational contraction dumps an immense amount of energy into the material. But as the core temperature exceeds 10 billion degrees Kelvin, the photons that are produced (black body radiation) are of such short wavelength (Wien's law), they are able to interact with and destroy the iron nuclei:

$$Fe^{56} + \text{high energy photon} \rightarrow 14\ He^{4}$$

The **photodissociation** of iron rapidly absorbs the energy released by gravitation, and the collapse is accelerated. When compressed enough, electrons are pressed into nuclei and helium ceases to exist:

$$2e^{-} + He^{4} + \text{energy} \rightarrow 4n + 2\nu$$

This reaction also cools the core not only because the conversion of protons to neutrons requires energy to occur, but because the neutrinos (ν) carry away additional energy, thus accelerating the collapse. The loss of electrons that contributed to supporting the core through their degeneracy further aids the collapse.

If the collapsing core is not too big, **neutronization** of the material can stop the collapse. Like electrons, at high enough density neutrons can exert a **neutron degeneracy pressure.** The neutrons find themselves in an overcompressed state. The core thus rebounds almost instantly, sending a shock wave outward into the star.

Exterior to the collapsing core, stellar layers still rich with light elements have also been falling inward. The shock wave from the rebounding core reverses their infall and blasts this material toward the surface. This expansion is augmented by absorption of energy from the neutrinos pouring out of the core (normally neutrinos do not interact, but here absorption of energy is significant as the matter immediately outside the core is incredibly dense and the production of neutrons has produced a very high neutrino flux). These layers heat up and runaway thermonuclear reactions ensue. With immense numbers of neutrons mixed into the material from the core, the full gamut of chemical elements is produced. The core collapses and initiation of the **type II supernova** explosion is rapid, occurring on a time scale of minutes.

As observed from outside, the inner core region is revealed when, a few hours after initiation of core collapse, the shock wave reaches the exterior of the star. The outer envelope is blown away at thousands of kilometers per second, and the star, originally very bright, increases by 100,000 to 1,000,000 times in luminosity (see Figure 12-4). Several solar masses of material may be blown away, including large quantities of newly formed heavy elements. Some 14 supernovae of this type (Type II) have been observed in the Milky Way Galaxy over the last 2,000 years, mostly by Chinese astronomers. These include SN 1054, which produced the Crab Nebula, visible 23 days in broad daylight; SN 1572, observed by Tycho Brahe; and SN 1604, observed by Kepler. SN 1987 in the Large Magellanic Cloud was the last naked-eye supernova. Other supernovae remnants, the expanding clouds of heavy element rich gas, are the Vela Nebula and the Cygnus Loop. This type of supernova is seen in other galaxies in association with young stellar populations. This is consistent with Type II supernovae precursors being young, evolved high-mass stars.

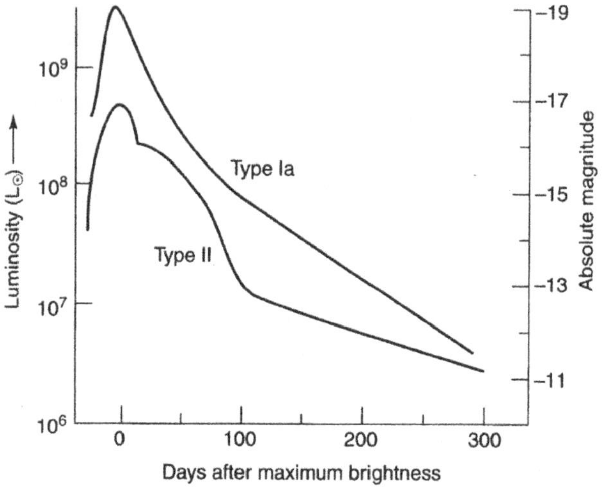

Figure 12-4

Type I and Type II supernovae light curves.

Neutron Stars (Pulsars)

If the collapsing core in a supernova explosion is less than about three solar masses, it can achieve a stable state with neutron pressure in balance against gravity. The result is a very compact object, a **neutron star,** with a radius of about 10 km and an extreme density of around 5×10^{14} g/cm^3—at the surface, a 1 mm grain of sand would weigh 200,000 tons. During the collapse, angular momentum conservation results in rapid rotation (see Chapter 4), many times per second initially, and conservation of magnetic field lines produces a magnetic field billions of times stronger than a normal star. The interior temperature is on the order of a billion degrees, and the neutrons act as a fluid there. A much cooler, thin, solid crust overlies this interior. Its very small surface area, however, results in an extremely low luminosity. In fact, astronomers have not yet detected the thermal radiation coming directly from the surface of a neutron star, but these objects are observable in another fashion.

Pulsars, stars observed to emit radiation in precisely separated pulses, were discovered in 1967. The first to be identified is coincident in position with the central stellar remnant in the Crab Nebula. Pulsars were quickly matched with the hypothetical neutron stars predicted in the 1930s. The pulses of radiation are due to a lighthouse beaming effect. The rapid rotation (the Crab pulsar rotates 30 times per second) carries the star's magnetic field around it, but at a radius not far from the star, the magnetic field would be rotating at the speed of light in violation of the theory of special relativity. To avoid this difficulty, the magnetic field (which generally is tilted with respect to the star's rotational axis) is converted to electromagnetic radiation in the form of two lighthouse beams directed radially outward along the magnetic field. An observer can detect a pulse of radiation every time a light beam passes by. Ultimately, therefore, it is the rotation of the star that is the energy source for the pulses and for the radiation that keeps the surrounding supernova nebula excited. For the Crab pulsar, this is about 100,000 times the solar luminosity. As rotational energy is lost, the star slows.

Unlike normal stars, neutron stars have a solid surface, with the neutrons locked into a crystalline lattice. As these stars radiate away energy, the crust slows its rotation. Observationally, the pulses are seen to be slowing down at a rate in agreement with the measured energy emission. But the fluid interior does not slow down. At some point, the disparity between their rotations results in an abrupt speedup of the crust, with an instantaneous decrease (a **glitch**) in the period of the pulses that are produced by the lighthouse beaming. In August 1998, a readjustment of this phenomenon in a distant neutron star apparently split open its outer crust, revealing the billion degree interior. This produced a significant flux of X-radiation, which momentarily bathed Earth, but fortunately for life on the planet's surface, was absorbed by the atmosphere.

The behavior of neutron stars in binary systems is analogous to binaries containing a white dwarf companion. Mass transfer can occur and form an **accretion disk** around the neutron star. Heated by the neutron star, this disk is hot enough to emit X rays. A number of **X-ray binaries** are known. When hydrogen from the accretion disk accumulates on the surface of the neutron star, rapid conversion to helium may be initiated, producing a brief emission of X rays. **X-ray bursters** may repeat this process every few hours to days.

In exceptional cases, mass infall onto an old neutron star (a dormant pulsar) with transfer of angular momentum may result in a significant spin-up of the star. A renewed rapid rotation will reinitiate the beaming mechanism and produce an extremely short period **millisecond pulsar.** Under other circumstances, the intense X-ray flux from a pulsar can actually heat the outer layers of a companion to the extent that this material escapes. Ultimately, the companion star may be completely vaporized.

Black Holes and Binary X Ray Sources

What happens if star can't get rid of enough mass in a supernova explosion to produce a remnant neutron core below three solar masses (below which only can neutrons produce enough pressure to counteract gravity); or if the collapse of the core is so dramatic as to smash through the neutron pressure barrier? When an object of mass M has radial size less than $R = 2GM/c^2$ (the **Schwartzschild radius;** 3 kilometers for a mass of 1 solar mass), then the surface gravitation becomes so intense that not even light may escape; the object disappears from view. Although not visible in any form of electromagnetic radiation, the object's gravitational field would still be felt in the surrounding space. Such a **black hole** could be detected by its gravitational influence on other objects.

Evidence for such collapsed objects appears to exist in the form of **binary x-ray systems.** Here a compact object may accrete material from its companion that is swelling to become a red giant star. As this material falls in toward the compact star, angular momentum conservation produces a rapidly rotating accretion disk near the compact star. The energy released from infall of additional matter and its collision with this accretion disk appears in the form of X-rays, gamma rays, and other energetic photons. Application of Kepler's Third Law to the observed orbital motion of the visible companion in several X ray sources (for example, Cygnus X-1) suggests that the masses of the unseen companions are too large to be any kind of known star; thus presumably the unseen stars are black holes.

In summary, objects termed stars may represent a wide variety of physical conditions, as shown in Table 12-1 and Figure 12-5:

Table 12-1: A Comparison of Stellar Properties

Type of Star	*Radius*	*Balance against Gravity*	*Structure*	*Mass/M_o*
Main Sequence	~solar (700,000 km)	Gas pressure	Gaseous	0.08–80
Giants, supergiants	10 – 400 solar	Gas pressure	Gaseous	0.5–80
White dwarfs	~Earth (6,400 km)	Electron degeneracy pressure	Solid	< 1.4
Neutron stars	~10 km	Neutron degeneracy pressure	Solid	< 3.0
Black holes	$<2GM/c^2$	None	None	–3.0

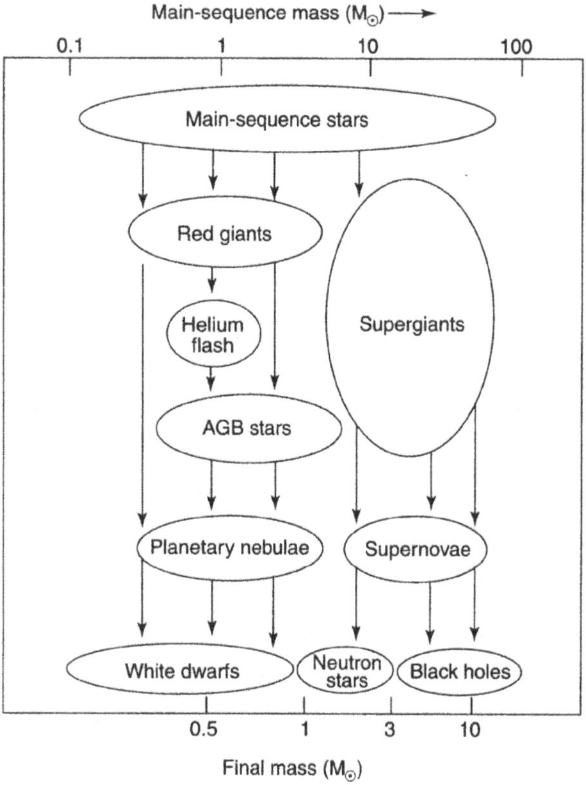

Figure 12-5

Summary of stellar evolution.

CHAPTER 13
THE MILKY WAY GALAXY

The **Milky Way Galaxy,** or simply the Galaxy, is a typical example of a **galaxy,** a large, independent system of stars, star clusters, and interstellar material. By studying the Milky Way, we can better understand galaxies as a whole. The Milky Way can be identified as a spiral galaxy because of the pinwheel-shaped interstellar material that traces out a spiral pattern in the plane of the Galaxy. The Galaxy is actually made up of two distinctly different spatial distributions of stars, one of which forms a flat disk and the other a centrally condensed, slightly flattened sphere that surrounds the disk.

Interstellar Matter

About 3 percent of the mass of the Milky Way exists in the form of **interstellar matter,** or diffuse material that floats between the stars. These extremely thin clouds of dust and gas average about 1 atom (mostly hydrogen) per cubic centimeter compared to air, which has a standard pressure and temperature of 10^{22} particles/cm^3.

The role of interstellar matter in astronomy is threefold. First, it provides the source of material for the formation of new stars. In turn, the material is partially replenished by mass loss from aging stars. Second, direct observation of interstellar matter provides scientists with information about the Galaxy that complements evidence obtained from studying stars. Finally, the presence of interstellar material along the line of sight to an object may dramatically affect the observation of the light coming from that object.

Gas

The two components of the interstellar material—gas and dust—must be clearly distinguished. Gas and dust are not only detected in different ways, but their distinct characteristics also cause them to affect observations of other objects differently. Gas, which makes up about 98 percent of the interstellar material, consists of individual atoms, primarily hydrogen and helium, and in dense regions there may also be molecules (more than 100 kinds of simple molecules, for example, water, formaldehyde, and alcohol, composed of up to 14 atoms have been identified so far). Gas has little effect on the observation of objects seen through it; it will add a few absorption lines to the spectrum of the light detected by the observer, but overall, **interstellar absorption lines** added to the spectrum of a distant star have a negligible effect on the star's observed brightness and color. The 21-cm wavelength of neutral hydrogen (HI) is especially important for studying of the Galaxy because this long wavelength passes through the dust without being absorbed. If it were not for this 21-cm radiation, most of the Galaxy could not be observed and studied by astronomers. Gas produces detectable, visible emission lines only in relatively dense, hot regions around hot stars.

Dust

The remaining 2 percent of the interstellar material is dust. Dust can be detected by the thermal radiation it produces, characteristic of its temperature in interstellar space (typically around 100 K). Its effect on the observation of other objects, however, cannot be ignored. Dust absorbs light that passes through it and scatters that light into other directions, thus dimming objects located in or behind it. In the plane of the Galaxy, dust effectively prohibits visually observing stars beyond about 3 Kpc (10,000 light-years) distance. The absorption is greatest for short-wavelength blue light; thus, the dust also makes distant objects appear redder (see Figure 13-1).

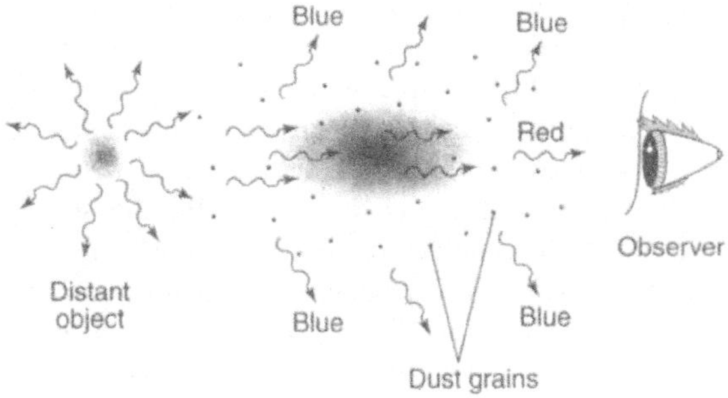

Figure 13-1

Interstellar absorption. Blue light from a distant object is preferentially scattered by dust along the line of sight, hence the object appears both redder and dimmer.

Interstellar Nebulae

The existence of interstellar material is most obvious where it is densest, in the form of clouds or **nebulae.** They form three basic types, those that emit their own radiation, those that are identified by absorption of the light from distant stars, and those that shine by reflected star light.

Emission nebulae (HII regions)

Emission nebulae, or nebulae that glow with their own light, exist where relatively dense interstellar material is found in the vicinity of hot stars. Such material may be left over from star formation (for example, the Orion nebula, which encloses several hundred young O and B stars) or from stellar mass loss in supernovae, planetary nebulae, or novae. Ultraviolet light from the hot stars is absorbed by the gas, ionizing the hydrogen (also termed HII, hence the alternative name **HII regions**) at a high temperature, around 10,000 K. As ionized hydrogen recombines with electrons, hydrogen emission lines in the visible part of the spectrum are produced. The net effect is conversion of stellar ultraviolet radiation into emission lines of hydrogen, primarily the reddish Hα line in hydrogen's Balmer series. Hence, emission nebulae have a characteristic red or pinkish color. The extent of the ionization around a star depends upon its ultraviolet output. An O5 star with a surface temperature of 50,000 K produces sufficient ultraviolet radiation to ionize hydrogen out to a distance of several parsecs (depending on the density of the gas), whereas a cooler star like the Sun is able to ionize surrounding hydrogen only within the confines of the solar system (~40 AU).

Absorption nebulae (dark nebulae)

If no hot stars are present, the interstellar gas will be cold, and dust may condense from the heavier elements. Interstellar dust has a size and structure similar to soot particles, and in the densest regions the dust may completely obscure the light of stars behind it, producing an **absorption nebula,** or **dark nebula.** Typical dust cloud sizes are about 10 pc, with representative masses of about 50 solar masses. Within these dusty regions, globules of denser material (Bok globules, of 0.1-1 pc diameter) may be found where self-gravity has led to a collapse of the interstellar material, leading to eventual star formation.

Reflection nebulae

The obscuring effect of dust is primarily the result of scattering or reflecting light into different directions. A dust cloud near stars but not close enough to destroy the dust through heat or ultraviolet radiation will preferentially reflect the stars' blue light; hence a bluish-looking nebula can be observed (the blueness of the daytime sky is produced in this same manner from scattering sunlight). The bluish wisps seen in the Pleiades star cluster are **reflection nebulae.**

Because young, hot stars are intermixed with older, cooler stars, both of which may be associated with interstellar material, it is not uncommon to find complicated structures in which all three types of nebulae are seen in close proximity.

Reddening and extinction of starlight

In addition to the obvious interstellar material, there is also a general interstellar dust that pervades all regions of space. This dust has the same effects as in the denser regions—it dims and reddens the light of stars. Nearby stars are relatively unaffected, but more distant stars are progressively fainter, not only because of the distance effect but also because of the dimming (technically termed **extinction**) of their starlight by the dust. Again because the dust affects blue light preferentially, distant stars appear progressively redder than they would if the dust were not present. In order to obtain correct measurements of stellar luminosities and colors, all observations must be corrected for this interstellar extinction and **reddening.**

The diffuse dust effectively hides stars that are more distant than 3 kpc in the galactic plane. Prior to 1920, however, astronomers did not recognize the existence of this dust. As a result, counts of stars in all directions led to the erroneous conclusion that the Sun was at the center of a stellar distribution—the so-called **Kapteyn universe**—whose equatorial plane measured roughly 17,000 pc in diameter and approximately 4,000 pc perpendicularly. With the recognition that dust hides the most distant stars, astronomers realized that this conclusion was erroneous, and they discarded it.

Interstellar absorption lines

Interstellar absorption lines are superimposed on the absorption features already present in stellar spectra. They are best studied by observation of O and B stars, which have the fewest intrinsic absorption lines, or in the spectra of unresolved binary stars. In these binary star spectra, pairs of stellar absorption lines (one from each star) will oscillate in wavelength as the stars alternatively move toward or away from observer (the Doppler effect), but non-stellar absorption lines produced along the line of sight will have fixed wavelengths. Interstellar absorption lines are also much narrower than stellar features because the gas temperature is much colder. Features attributable to elements such as H, He, Na, K, Ca, Be, Fe, Ca, and Ti and molecules of CN and CH have been identified and show that the interstellar chemistry is like that of stars. Often multiple interstellar absorption features are found, suggesting that the interstellar material is not uniformly distributed, but has a wispy cloudlike structure.

Interstellar molecules and giant molecular clouds

In dense regions where dust shields interstellar material from the ultraviolet radiation of surrounding stars, molecules can form. Scientists believe that the chemical reactions that produce molecules occur on the surfaces of dust grains from which the molecules subsequently escape. Such molecules (some 120 or so have been identified) are made up of the most common elements (hydrogen, oxygen, carbon, and nitrogen). Molecules, unlike individual gas atoms, have vibrational and rotational states that allow radiation in the infrared and radio part of the spectrum. Such long wavelengths are unaffected by the interstellar dust and thus can be detected coming from within these dense regions, allowing astronomers to study the structures and the dynamics of dense interstellar clouds. The detection of such radiation also led to the discovery of the **giant molecular clouds,** huge assemblages of up to 10^6 solar masses of gas that otherwise are invisible to observation.

21-cm radio radiation from neutral hydrogen

The most important radio radiation coming from the interstellar material is the **21-centimeter radiation** (a wavelength that corresponds to a frequency of 1,421 MHz) coming from hydrogen, the most common element. This long wavelength is not affected by dust and can be observed coming from every region in the Galaxy. Radio telescopes are extremely sensitive, so even very weak signals may be detected.

Star Clusters

During the gravitational collapse of large clumps of interstellar material, the material may often fragment into smaller pieces. From these smaller pieces of interstellar matter, multiple star systems or star clusters may form whose mutual gravitation makes them stable against dispersal for reasonably long periods of time. Being of the same age, stars of different masses within the clusters will be found in different evolutionary stages, providing a major test of evolutionary theory. Young clusters allow astronomers to test theories on how stellar birth occurs, and old clusters with white dwarfs enable them to test theories of the last stages of stellar evolution. Star clusters are also important because they enable astronomers to place stars of different types in their proper positions within the HR diagram. Moreover, because star clusters are relatively compact objects that are easily recognized against the more uniform distribution of background or foreground stars, they are easy to discover. Consequently, star clusters have played decisive roles in mapping the overall structure and size of the Galaxy, identifying regions of the most recent star formation, establishing the age of the Galaxy, and providing information about its evolution.

The two types of clusters, **open clusters** and **globular clusters,** have distinctly different characteristics. Astronomers also recognize a third type of stellar assemblage, **associations,** whose stars are coeval in formation, but that over time have dispersed over a much wider region in space.

Open clusters

Open clusters are the smaller clusters of stars found in the Galaxy, both in terms of number of member stars (typically 10–500 stars) and size (3–10 pc in diameter). These clusters are located in the plane (or disk) of the Milky Way where they have the same motion as the Sun around the center of the Galaxy. As a result, their observed velocities are small with respect to the Sun. Such a small grouping of stars is obvious against the general background (or foreground) of stars, hence they are easily identified; some 4,000–5,000 are listed in modern catalogs. Open cluster ages are predominantly young, and they are often associated with interstellar material by virtue of this youth. The youngest of these clusters have bright main sequence stars, blue supergiants, and sometimes a few variables, either young Cepheids or even younger T Tauri stars. Open clusters are not tightly bound, as their few stars give them a density of 0.1–10 stars/pc^3 and a gravitational self-attraction that is relatively weak. Few open clusters therefore survive to old age; the oldest identified clusters are about 8×10^9 years old. Chemically, their spectra show compositions similar to that of the Sun (that is, a solar abundance of about 98 percent hydrogen and helium, and 2 percent heavy elements). All these properties represent those characteristic of Population I objects in the Galaxy (see Stellar Populations, later in this chapter).

Associations

Associations are loose groupings of stars. About 80 associations are known, and they are related to open clusters in that they are young objects located in the galactic disk and they have a common formation. Unlike open clusters, however, associations are not gravitationally bound, and the stars in them are slowly moving away from each other. Associations are identified on the basis of the similarity of their motions through space and other properties, although their few dozen to few hundred stars may be spread over several hundred parsecs of space (representative stellar density of $<0.01\ pc^{-3}$). They are typed as O, B, A, or T Associations on the basis of the types of stars in the group. Associations older than about 10 million years are not known because their stars have become too dispersed to be recognizable on the basis of their common properties. One prominent example of an association includes five of the seven stars in the Big Dipper, the bright star Sirius, and several others.

Globular clusters

In all their properties, the globular clusters (there are about 150 associated with the Milky Way Galaxy) are distinct from the open clusters. They are old clusters, from 12 to possibly 18 billion years old, and are certainly among the first objects to have formed in the Galaxy. Globular clusters are populous, with 10^4 to a few times 10^6 stars with diameters of up to a hundred or more parsecs and their central densities as tightly packed as 10–1000 stars/pc^3. Containing no young stars, they have many red giants and old variable stars such as RR Lyraes. In the Milky Way, these clusters are found distributed in a more or less spherical halo surrounding the Galaxy, with their greatest concentration toward the center of the Galaxy (in the direction of the constellation Sagittarius). Overall, the system of globular clusters shows little evidence of rotation, with orbital motions in randomly oriented and highly elliptical paths (predominantly radial motions). Relative to the Sun, these clusters have heavy element abundances much smaller than the Sun, hence are termed **metal-poor.** In sum, the properties (age, chemistry, spatial distribution, motions) illustrated by the globular clusters are characteristic of Population II stars in the Galaxy (see Stellar Populations, later in this chapter).

Structure of the Galaxy

Passing around the sky there is a broad region that is readily seen to be brighter than the rest of the night sky. It has been traced from the summer constellation Sagittarius northward through Cyngus into Perseus, then southward to Orion (winter sky) into Centaurus (Southern Hemisphere sky) then back northward into Sagittarius. Even a small telescope or pair of binoculars reveals this band to be bright because of the cumulative effect of millions of faint stars. This is the Milky Way. That it is due to myriads of faint stars distributed in a great circle about the position of the Sun shows the Galaxy's basic structure, the manner in which the stars and interstellar material that make up the Galaxy are distributed in space, is flat. This is the **plane** of the Galaxy, where the greater part of the stars and interstellar material exist. The brightest part of Milky Way, visible low on the southern horizon in the summer sky toward the constellation of Sagittarius, is bright because the star density increases in this direction. This is the direction to the center of the Galaxy, though starlight coming from the vast bulk of stars in this direction is invisible because of the absorption by the dust.

The distribution of dusty, absorption nebulae is very patchy, and there are "windows," directions passing close to the center in which there is relatively little absorption, that allow study of the distant stars. In these directions and elsewhere in the halo of the Galaxy, the distribution of RR Lyrae and other stars yields its density structure. In the same manner, the directions and distances to the globular clusters may be mapped in three dimensions. The clusters are concentrated in the direction of Sagittarius, and their density decreases outward, allowing astronomers to outline the outer structure of the Galaxy. From their distribution, the position of the densest part of the Galaxy, the center, may be determined. The galactocentric distance of the Sun is currently estimated as $R_{\odot} \approx 8$ Kpc (25,000 ly).

The brightest stars at the center of the Galaxy may also be studied using long wavelength infrared radiation. The total extent of the plane of the Galaxy can be deduced by analyzing observations of the 21-centimeter radiation of neutral hydrogen 360° around the plane. This analysis gives the size of the whole Galaxy as about 30,000 pc diameter (100,000 ly). Scans in 21-cm above and below the plane, together with observations of stars perpendicular to the plane, give a total thickness of about 500 pc (1,600 ly), with half the gas mass within 110 pc (360 ly) of the center of the plane. Radio studies also reveal that the fundamental plane of the Galaxy is warped, like a fedora hat, with the brim pushed up on one side and down on the other (see Figure 13-3). It is bent down on Sun side of the Galaxy and up on the opposite side, due to a gravitational resonance with the Magellanic Clouds, which move in an orbit about the Milky Way.

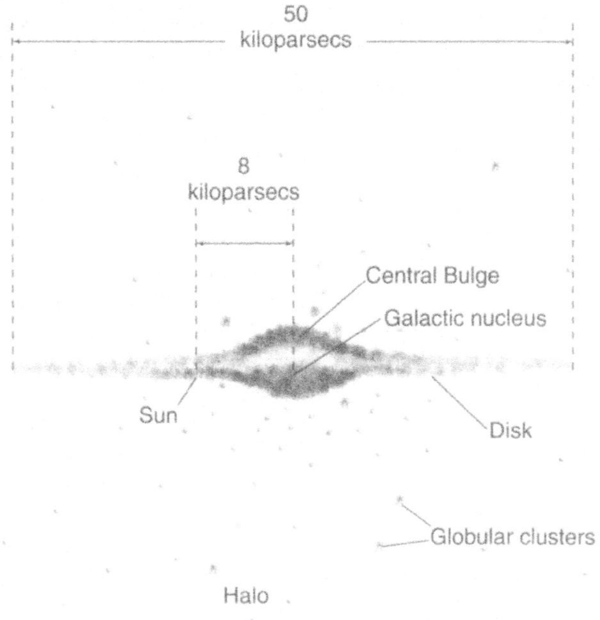

Figure 13-2

An external view of the Milky Way, looking edge-on or sideways into the disk.

While the greater part of the mass of the Milky Way lies in the relatively thin, circularly symmetric plane or disk, there are three other recognized components of the Galaxy, each marked by distinct patterns of spatial distribution, motions, and stellar types. These are the halo, nucleus, and corona.

Disk

The **disk** consists of those stars distributed in the thin, rotating, circularly symmetric plane that has an approximate diameter of 30,000 pc (100,000 ly) and a thickness of about 400 to 500 pc (1,300 to 1,600 ly). Most disk stars are relatively old, although the disk is also the site of present star formation as evidenced by the young open clusters and associations. The estimated present conversion rate of interstellar material to new stars is only about 1 solar mass per year. The Sun is a disk star about 8 kpc (25,000 ly) from center. All these stars, old to young, are fairly homogeneous in their chemical composition, which is similar to that of the Sun.

The disk also contains essentially all the Galaxy's content of interstellar material, but the gas and dust are concentrated to a much thinner thickness than the stars; half the interstellar material is within about 25 pc (80 ly) of the central plane. Within the interstellar material, denser regions contract to form new stars. In the local region of the disk, the position of young O and B stars, young open clusters, young Cepheid variables, and HII regions associated with recent star formation reveal that star formation does not occur randomly in the plane but in a **spiral pattern** analogous to the **spiral arms** found in other disk galaxies.

The disk of the Galaxy is in **dynamical equilibrium,** with the inward pull of gravity balanced by motion in circular orbits. The disk is fairly rapidly rotating with a uniform velocity about 220 km. Over most of the radial extent of the disk, this circular velocity is reasonably independent of the distance outward from the center of the Galaxy.

Halo and bulge

Some stars and star clusters (globular clusters) form the **halo** component of the Galaxy. They surround and interpenetrate the disk, and are thinly distributed in a more or less spherical (or spheroidal) shape symmetrically around the center of the Milky Way. The halo is traced out to about 100,000 pc (325,000 ly), but there is no sharp edge to the Galaxy; the density of stars simply fades away until they are no longer detectable. The halo's greatest concentration is at its center, where the cumulative light of its stars becomes comparable to that of the disk stars. This region is called the (nuclear) **bulge** of the Galaxy; its spatial distribution is somewhat more flattened than the whole halo. There is also evidence that the stars in the bulge have slightly greater abundances of heavy elements than stars at greater distances from the center of the Galaxy.

The halo stars consist of old, faint, red main sequence stars or old, red giant stars, considered to be among the first stars to have formed in the Galaxy. Their distribution in space and their extremely elongated orbits around the center of the Galaxy suggest that they were formed during one of the Galaxy's initial collapse phases. Forming before there had been significant thermonuclear processing of materials in the cores of stars, these stars came from interstellar matter with few heavy elements. As a result, they are metal poor. At the time of their formation, conditions also supported the formation of star clusters that had about 10^6 solar masses of material, the globular clusters. Today there exists no interstellar medium of any consequence in the halo and hence no current star formation there. The lack of dust in the halo means that this part of the Galaxy is transparent, making observation of the rest of the universe possible.

Halo stars can easily be discovered by proper motion studies. In extreme cases, these stars have motions nearly radial to the center of the Galaxy—hence at right angles to the circular motion of the Sun. Their net relative motion to the Sun therefore is large, and they are discovered as **high-velocity stars,** although their true space velocities are not necessarily great. Detailed study of the motions of distant halo stars and the globular clusters shows that the net rotation of the halo is small. Random motions of the halo stars prevent the halo from collapsing under the effect of the gravity of the whole Galaxy.

Nucleus

The **nucleus** is considered to be a distinct component of the Galaxy. It is not only the central region of the Galaxy where the densest distribution of stars (about 50,000 stars per cubic parsec compared to about 1 star per cubic parsec in the vicinity of the Sun) of both the halo and disk occurs, but it is also the site of violent and energetic activity. The very center of the Galaxy harbors objects or phenomena that are not found elsewhere in the Galaxy. This is evidenced by a high flux of infrared, radio, and extremely short wavelength gamma radiation coming from the center, a specific infrared source known as Sagittarius A. Infrared emissions in this region show that a high density of cooler stars exists there, in excess of what would be expected from extrapolating the normal distribution of halo and disk stars to the center.

The nucleus is also exceptionally bright in radio radiation produced by the interaction of high-velocity charged particles with a weak magnetic field (**synchrotron radiation**). Of greater significance is the variable emission of gamma rays, particularly at an energy of 0.5 MeV. This gamma-ray emission line has only one source—the mutual annihilation of electrons with anti-electrons, or positrons, the source of which in the center has yet to be identified. Theoretical attempts to explain these phenomena suggest a total mass involved of 10^6–10^7 solar masses in a region perhaps a few parsecs in diameter. This could be in the form of a single object, a **massive black hole;** similar massive objects appear to exist in the centers of other galaxies that show energetic nuclei. By the standards of such active galaxies, however, the nucleus of the Milky Way is a quiet place, although interpretations of the observed radiation suggest the existence of huge clouds of warm dust, rings of molecular gas, and other complex features.

Exterior to the halo

The gravitational influence of the Galaxy extends to an even greater distance of about 500,000 pc (1,650,000 ly) (the late astronomer Bart Bok suggested this region could be called the corona of the Galaxy).

In this volume there appears to be an excess of **dwarf galaxies** associated with the Milky Way, drawn into its proximity by its large gravitational pull. This includes the **Magellanic Clouds,** which lie in the debris of the **Magellanic Stream.** The Magellanic Stream consists of a band of hydrogen gas and other materials that extends around the Galaxy, marking the orbital path of these companion galaxies. The tidal gravitational field of the Galaxy apparently is ripping them apart, a process that will be completed in the next two to three billion years. This **galactic cannibalism,** the destruction of small galaxies, and the accretion of their stars and gas into a larger galactic object likely has happened in the past, perhaps many times. A second, small companion galaxy in the direction of Sagittarius (the Sagittarius galaxy) appears to be another victim of this process. Like the Magellanic Clouds, its stars and interstellar material will ultimately be incorporated into the body of the Milky Way. The total number of dwarf galaxies near the Milky Way is about a dozen and includes objects such as Leo I, Leo II, and Ursa Major. A similar cloud of dwarf galaxies exists about the Andromeda Galaxy.

Rotation curve of the Galaxy

An alternative means of studying the structure of the Galaxy, complementary to looking at the distribution of specific objects, is to deduce the total distribution of mass. This may be done by analyzing the **rotation curve,** or the circular velocity V(R) of the disk objects moving around the center of the Galaxy as a function of the distance R out from the center. A check on the accuracy of the deduced motion in the Galaxy is given by the rotation curves of similar galaxies, which would be expected to rotate in the same basic fashion. Like the Milky Way, the rotations of other galaxies show a linear increase of velocity near their centers rising to a maximum value and then becoming basically constant over the remainder of the disk.

Determination of V(R) from within the Galaxy is not as straightforward as measuring the rotation of another galaxy that is observed from outside. Observation of neighboring stars or of interstellar gas gives only *relative* motions. Thus, calculating the absolute solar velocity involves first looking at nearby galaxies and determining what direction the Sun appears to be moving in.

The Sun and its neighboring stars are found to be moving about the center of the Galaxy with a speed of 220 km/s in the direction of the northern constellation Cygnus, at a right angle to the direction towards the center. In the **galactic coordinate system** used by astronomers, this movement is toward a galactic longitude of 90°. Sweeping around the Galaxy in its plane, **galactic longitude** starts at 0° toward the center, increases to 90° in the direction of rotation (Cygnus), to 180° in the anti-center direction (Orion), to 270° in the direction from which the Sun moves (Centaurus), and finally to 360° when the direction of the center is again reached. Use of Doppler shifts and proper motions applied to stars near the sun provide some idea of the local rotation curve; nearby disk stars on average appear to move in circular orbits about the center with the same circular velocity as the Sun. The interstellar dust prevents study by optical techniques of the rest of the Galaxy; thus, the 21-centimeter radiation of neutral hydrogen must be used to determine its pattern of motion. Again, the Doppler Shift gives only a relative or line-of-sight velocity for the gas anywhere in the Galaxy, but knowledge of the solar velocity and geometry allows calculation of the velocity at other radii from the galactic center.

The rotation curve of the Galaxy shows that it does not rotate as a solid disk (velocity directly proportional to distance out from the rotational axis). Rather, the rotational velocity is more or less constant over most of the disk (see Figure 13-3). Viewed as a giant race course, this means that on average all stars move the same distance in a given amount of time, but because the circular paths of outer stars are larger than those closer to the center, the outer stars slip progressively behind the inner stars. This effect is called **differential rotation,** and it has significant effects on the distribution of star-forming regions; any large star-forming region will be sheared into a spiral arc. If the Galaxy rotated as a solid disk, there would be no differential rotation.

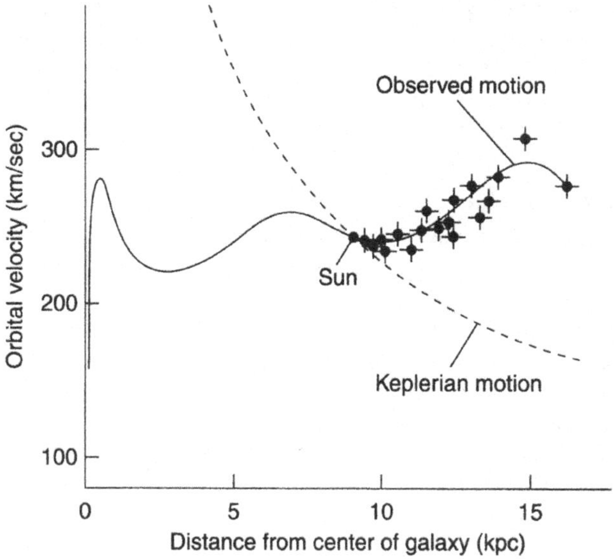

Figure 13-3

Rotation curve of the Galaxy. If the greatest portion of the mass of the Galaxy were concentrated at its center, then orbital motions would rapidly decrease with radius (dashed line) in the manner of the planetary motions about the Sun described by Kepler.

Stars, including the Sun, have small components of motion that deviate from a pure circular motion about the center of the Galaxy. This **peculiar motion** for the Sun is about 20 km/s, a small drift in the general direction of the bright summer star Vega. This results in an approximate 600 pc (1900 ly) in-and-out deviation from a true circular orbit as the Sun orbits the center of the Galaxy with a period of 225 million years. A second consequence is an oscillation, with a much shorter period of about 60 million years, up and down through the plane of the disk. In other words, the Sun moves up and down about four times during each trip around the center of the Galaxy. This oscillation has an amplitude of 75 pc (250 ly). At present, the Sun is 4 pc (13 ly) above the galactic plane, moving upward into the Galaxy's Northern Hemisphere.

Mass distribution

In one sense, the Galaxy is analogous to the solar system: The flatness is the result of the operation of the same physical laws. As the material of both contracted at their time of formation, conservation of angular momentum resulted in increased rotational velocities until a balance against gravity was achieved in an equatorial plane. Material above or below that plane continued to fall inwards until the mass distribution became flat. In specific detail, the mass distributions are very dissimilar. The mass of the Galaxy is distributed through a large volume of space, whereas the mass of the solar system is essentially only that of the Sun and is located at the center. The flat disk of the Galaxy implies that rotation plays the dominant role in the balance against gravitation, which, in turn, depends on the mass distribution. The mass M(R) as a function of radius R is determined by applying a modification of Kepler's Third Law to the rotation curve V(R), to obtain

$$M(R) \approx RV^2(R) / G$$

where G is the gravitational constant. Thus, astronomers can determine the mass structure of the Galaxy. Its total mass may be as great as 10^{12} solar masses.

Because the mass in the Galaxy is distributed over a large volume, the pattern of rotation differs from that in the solar system. For the planets, orbital velocities decrease with radial distance outward, $V(R) \propto R^{-1/2}$ (Keplerian motion); in the Galaxy, the circular velocity rises linearly $V(R) \propto R$ near the center, and then is relatively unchanging over the remainder of the disk, $V(R) \approx$ constant. This form of rotation curve implies a relatively constant mass density near the center; but further out, the density decreases inversely with the square of the radius.

The motions of the stars are also affected by the spatial distribution of the mass. The nature of Newtonian gravity is that a circularly or spherically symmetric mass distribution always exerts a force toward the center, but this force depends *only on that part of the mass that is closer to the center than the object* that feels the force. If a star moves outward in the Galaxy, it feels the gravitational force from a larger fraction of the total mass; when it moves closer to the center, less of the mass is exerting a force on the object. As a result, orbits of stars are not closed ellipses like those of the planets, but instead more closely resemble the patterns produced by a spirograph. Additionally, a planetary orbit is a flat plane; hence, if that orbit is inclined to the overall plane of the solar system, in one complete circuit about the Sun the planet moves once above and once below the solar system plane. A star, however, will oscillate up and down several times in one passage around the center of the Galaxy.

Spiral arm phenomenon

In the Galaxy, the mass structure of the disk is not perfectly smooth. Instead, there are regions in the disk where the density of stars is slightly larger than the average. In these same regions, the density of the interstellar material may be significantly larger. These density variations, or fluctuations, are not completely random; they show a global pattern of spirality, or spiral arms, within the disk (see Figure 13-4). Again the dust in our Galaxy is a problem; thus, spiral features easily studied in distant disk galaxies can give us insight to the pattern in the Milky Way. Stellar and nonstellar objects associated with the spiral arms can be mapped out only locally in our Galaxy, out to 3 kpc (10,000 ly) or so, because in regions of higher density of interstellar material, star formation occurs. In particular, the brightest O and B stars are indicative of the most recent star formation. They and other objects associated with recent star formation (emission regions, Cepheid variables, young star clusters) may be used as optical tracers of the spiral arm pattern. Analysis of 21-centimeter observations is more difficult, but suggests that coincident with young stellar objects are the denser regions of interstellar material.

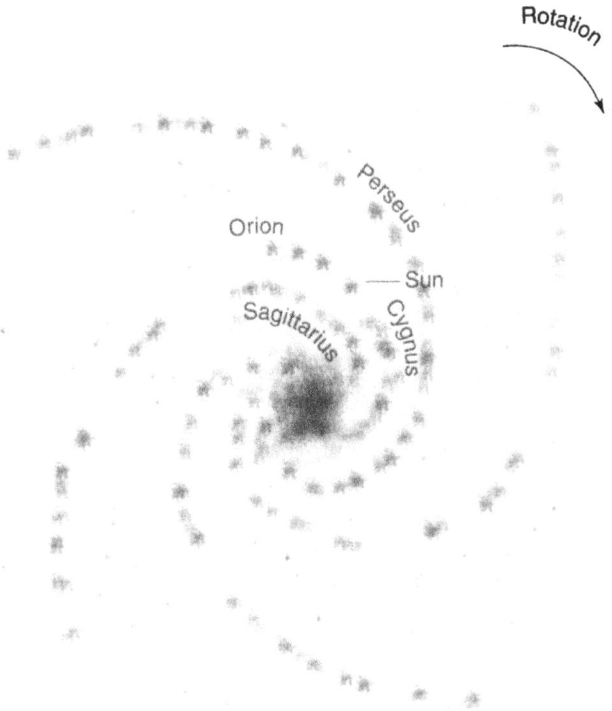

Figure 13-4

A schematic interpretation of the spiral features in the disk of the Milky Way Galaxy. The various spiral arms are named after the constellations in which directions their brightest features are observed.

To have a pattern of compression (higher density) and rarefaction (lower density) in the spiral arm pattern that exists over the whole disk of a galaxy requires energy, in the same manner that the sound produced when a person speaks requires energy. Both phenomena are examples of wave phenomena. A sound wave is a pattern of alternate compression and rarefaction in air molecules. Like any wave phenomena, the energy that is responsible for the wave will dissipate into random motions, and the wave pattern should die away in a relatively short period of time.

The density wave that passes through the disk of the Galaxy can be better related to the density waves that are found on freeways. At times, any given driver will be in the midst of "traffic," but at other times, he or she will seem to be the only driver on the road. Physically, these waves are the result of two factors. First, not all automobiles are driven at the same speed. There are slower and faster drivers. Second, congestion occurs because there are a limited number of lanes for the traffic flow. Faster drivers come up from behind and are delayed as they weave from lane to lane in their effort to get through to the head of the pack and resume their higher speed. They then can rush ahead, only to get caught up in the next pattern of congestion. Slower drivers get left behind until the next traffic wave catches up to them. Seen from a helicopter, a wave of alternatively denser and thinner distributions of cars is traveling down the highway; those cars in the dense regions, however, change as the faster cars move through and the slower ones drift behind.

In the Galaxy, the dynamics are slightly different in that the "highway" is a circulation about a galactic center, and the congestion is due to the stronger gravity in regions with larger numbers of stars. The **spiral density wave theory** begins by postulating the existence of a spirally structured pattern of density enhancement in a galactic disk. In the regions of extra density, the extra gravity affects motions and causes the gas and stars to "pile up" momentarily in these spirally shaped regions. Once the stars have passed through the spiral arm, they can move slightly faster until they catch up to the next spiral arm where they again will be momentarily delayed. The gas particles, being much less massive than the stars, are significantly more affected by the excess gravity and can be compressed to five times the average density of the interstellar matter in the disk. This compression is enough to trigger star formation; the newly formed luminosity O and B stars and their associated emission regions thus light up the regions of the spiral arms. The theory very successfully shows that a spiral density enhancement in the form of two well-formed spiral arms, a so-called **Grand Design,** is self-sustaining for several rotations of a galaxy. In the Milky Way, the expected flow pattern in stellar motions due to acceleration by the gravity of the spiral arms, superimposed on the overall circular motion about the center of the Galaxy, has been observed.

The evidence for the excitation of the wave in the first place should be evident because the lifetime of such a wave is rather short (a few galaxy rotation periods). In fact, a Grand Design spiral galaxy is generally accompanied by a companion galaxy whose recent close passage by the larger galaxy gave the gravitational stimulus to produce the density wave.

Not all galaxies show a distinct, two-armed spiral pattern. In fact, the majority of disk galaxies show numerous arc-like features, apparent fragments of spiral features that are referred to as **flocculent galaxies.** Each arc represents a region lit up by the bright stars of recent star formation and are explained by the **stochastic self-propagating star formation theory.** Given an initial collapse of interstellar gas into a group of stars, in due course a massive star will undergo a supernova explosion. Shock waves moving outward then push the ambient interstellar material into denser condensations and can trigger a next generation of new stars. If there are new massive stars, there will be subsequent supernovae, and the process repeats (the self-propagating aspect). This cycle continues until the interstellar gas is depleted, or until by chance no new massive stars form (this is the random, or stochastic, aspect of this theory). During the existence of a wave of star formation moving outwards from some original position, however, the growing region of star formation is affected by differential rotation in the disk; the outer part of the star-forming region lags behind the inner part. The region of star formation is therefore smeared into a spiral arc, as would be all other growing, star-forming regions elsewhere in the disk; but there would be no grand design.

Origin and Evolution of the Galaxy

The conventional picture of the formation of the Galaxy was developed to explain the spatial distribution, motions, and chemical properties of the stars that are found in the Galaxy. Initially, two distinct groups of stars, or stellar populations, were recognized by their very different properties.

Stellar populations

The most distinct component of what was defined as **Population I** are the open clusters and associations whose brightest stars are the luminous, blue, and young O and B stars. Such clusters are often associated with the interstellar material out of which these stars recently formed. On the other hand, the globular clusters representing **Population II** are very different stars, containing no O and B stars or gas and dust, but filled with old red giant stars.

Population clusters' differences include more factors than simply their time of formation, however, because they differ significantly in their space distribution and motions. Open clusters, for example, are located in the disk and have small velocities relative to the Sun. On the other hand, globular clusters are located in a spheroidal halo concentrated in the galactic center and are generally observed to have large velocities relative to the Sun. Chemically, the open clusters are similar to the Sun, possessing a fraction of heavy elements that range from about one-third to twice the solar abundance. In contrast, the globular clusters are relatively metal poor, with heavy element abundances between 0.001 and 0.5 times solar abundance.

The characteristics of these two classes of star clusters are indicative of the overall characteristics of other stars in the halo and disk. Astronomers now understand that their properties characterize not two truly distinct populations, but rather the extremes of a continuous distribution of stellar types, whose properties range from the spheroidally distributed, metal-poor stars to those metal-rich stars confined to a very thin plane in the disk. Stars with an even smaller content of heavy elements are the nearly pure hydrogen-helium stars, which have been discovered and represent the once hypothetical **Population III,** the first generation of stars in the Galaxy.

In the standard model for the formation of the Galaxy, the motions of the stars and their spatial distribution as observed at the present time reflect the conditions during the phase in which they formed. This is postulated to have begun very early in the history of the universe when some 10^{12} solar masses of primordial hydrogen and helium gas began to collapse under its own self-gravitation. The first stars to form would have been pure hydrogen and helium; but rapid stellar evolution of massive stars and their subsequent supernovae would have "polluted" the remaining interstellar material with heavy elements. The next generation of stars (Population II) would have had a small fraction of heavy elements, but their stellar evolution would have lead to ever greater additions to the heavy element content of the interstellar medium. The earliest generations of stars (including the globular clusters) forming during the collapse phase retain a memory of this in their nearly radial orbits. The gas, still the largest fraction of the mass of the Galaxy at this era, progressively flattened into a rotating disk because of angular momentum conservation, with each successive generation of stars being marked by a spatial distribution indicative of the gas from which they formed. During the flattening, collisions between the gas particles regularized motions until only circular motions survived. This process has continued to the present day, with the remaining interstellar gas, now significantly enriched with metals, in a very thin plane, in which the most recent Population I stars continue to form.

Many aspects of the present Galaxy, however, suggest that the true process of formation has been more complicated. A major alternative theory suggests that the collapse of pre-existing gaseous material again formed very flat disks, smaller galaxies similar to, but not quite the same, as the spiral galaxies identified in the present universe. Assemblages of these proto-spiral galaxies merged over time to form the large Milky Way Galaxy of today. Regardless of which process best describes the past of the Galaxy, it is apparent that the capture or cannibalism of other smaller galaxies has played a significant role in the history of the Galaxy.

CHAPTER 14
GALAXIES

Types of Galaxies and Their Classifications

The identification of other **galaxies,** or independent stellar systems, goes back to 1924, when Edwin Hubble found Cepheid variables in the nearby galaxies Messier 33 and Messier 31. Application of the Period-Luminosity Relation for these variables in the Magellanic Clouds established that these were objects outside the confines of the Milky Way and of sizes comparable to it. More thorough studies show that galaxies exhibit a wide range of properties, from **giant galaxies** of 10^{13} solar masses and sizes over 150,000 pc in diameter to **dwarf galaxies** of 10^{6} solar masses and about 1,000 pc in size. Dwarf galaxies are by far the most numerous. It is estimated that about 100 billion galaxies in the universe are observable with existing telescopes.

Hubble quickly realized that the vast majority of galaxies have only a small number of shapes. A classification based on their optical appearance or morphology is limited to four fundamental types of galaxies—ellipticals, spirals, irregulars, and S0s. The Hubble classification has proven to be immensely valuable to the study of galaxies. Although originally based solely on optical appearance, appearance is also closely correlated with other physical properties of galaxies.

Elliptical (E) galaxies

Ellipticals (also sometimes called *early-type galaxies*) were so named because they look like elliptical blobs of light. In general, they show no obvious structural features other than a smooth concentration of light to the center. The decrease in surface brightness with distance may be expressed in different ways, but one reasonable approximation is $I(r) = I_{\odot}/(a + r)^2$ where $I_{\odot}$ is a central brightness,

r is the distance from the center, and a is a distance at which the brightness is one-quarter of that in the center. In other words, the brightness roughly falls off as the inverse square of the distance from the center of the galaxy.

Many ellipticals are round, but others are noticeably elongated or flattened. If the long axis is measured to have a dimension of a and the perpendicular short axis is measured as b, then an ellipticity can be defined as $\varepsilon = 10\,(1 - b/a)$; rounded to the nearest unit, ε is used as a subtype to distinguish between ellipticals (E) with different shapes. An E0 is a round galaxy, whereas an E6 is a rather flattened system (but not a disk in the sense of a flat spiral galaxy) (See Figure 14-1). A serious problem with ellipticals, however, is determination of their real shape: A flat elliptical may look round if seen from above or below or face-on in the same manner that a dinner plate can look very different depending upon the position of the viewer.

Statistical studies suggest that the typical elliptical is moderately flattened; but this argument rests on an implicit assumption that ellipticals have an equatorial or circular symmetry, like a pumpkin (the technical description is an **oblate spheroid**). Such would be the case if the flattening were related to rotation, in the same sense that the equatorial bulge of a planet like Jupiter is produced by its rapid rotation. But ellipticals show only a slow rotation; the balance against gravitation is primarily accomplished by random (in and out) motions of the stars, not by rotation. Theoretical studies suggest that the true spatial distribution of stars in an elliptical is more similar to a bar-like structure (for example, like an eraser) known as a **tri-axial spheroid.**

Of all classes of galaxies, elliptical galaxies show the widest range of properties between the dwarf examples and the giant systems, with mass ranging from 10^6 to 10^{13} solar masses, sizes from 1 kpc to 150 kpc in diameter, and luminosities 10^6 to 10^{12} solar luminosities. Perhaps 70 percent of all galaxies are ellipticals, but the vast majority are dwarfs.

In terms of stellar content, ellipticals appear to contain no bright, young stars and, in fact, most show no evidence of recent star formation at all. But some ellipticals, especially those in the center of clusters, do show blue stars and a UV excess indicating recent star formation. With overall reddish colors, ellipticals were long considered to contain a single population of old stars with the brightest stars being red giants. These old stars, however, are not standard Population II stars as in the Milky Way Galaxy (see Chapter 13), because spectroscopic analysis shows that many of them have a metallicity like the Sun, or even a greater abundance of heavy elements. The past star formation history of an elliptical thus must be very different than that which occurred in the Galaxy. Ellipticals appear to be pure star systems, with virtually no interstellar material ($< 0.01\%$ of the total mass), although there are a few exceptions to this rule. This lack of interstellar matter poses a problem, because stars evolve and lose mass. Because ellipticals do not appear to be forming new stars that would get rid of such gas over the lifetime of an elliptical, about 2 percent of the mass would have been returned to the interstellar medium (assuming that one had 100 percent conversion of material into stars at the time of formation of the galaxy).

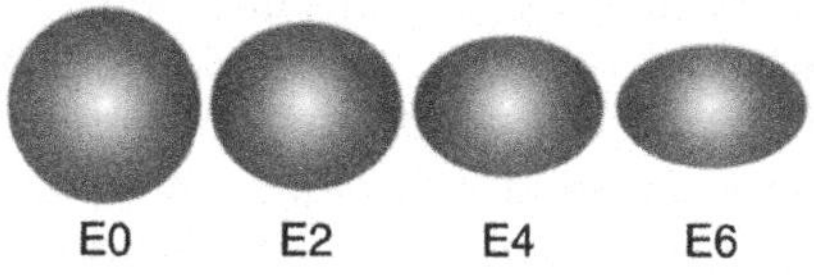

Figure 14-1

Elliptical galaxies.

Spiral (SA and SB) galaxies

About 15 percent of galaxies are **spirals,** flat galaxies with a central light concentration that show spiral arms in an outer disk. The central regions of spiral galaxies appear reddish and are composed of older Population II stars, such as those in the halo of the Milky Way

Galaxy. These stars are distributed in an almost spherical region around the center of a galaxy and exhibit little rotation. Their concentration toward the center produces the appearance of a central bulge in the light distribution. The outer disks of spirals appear bluish because of the presence of young, blue stars that have formed relatively recently out of the interstellar material. Redder stars are present in the arms as well, though they are not as bright and therefore contribute less to the brightness of the arms. The star formation is concentrated into the spiral arms that look brighter because of the exceptionally luminous O and B stars. In reality, the mass distribution in the disk is very smooth, with the spiral arm regions representing only a small density excess over the mean density (this is true even though the density enhancement for interstellar gas, a minor part of the total mass distribution, may be large). Circular motions predominate in the disk, and all other characteristics of the stars are typical of Population I objects like those of the Milky Way. The outer mass distribution (as implied by the distribution of light) is clearly different than that of the elliptical galaxies. Surface brightness in the disk decreases radially outward as $I(r) = I_{\odot} \exp(-r/a)$ where the length *a* represents a scale factor, a distance over which the brightness drops by a given amount.

Spiral galaxies range from intermediate to large galaxies, with masses in the range of 10^9 to 10^{12} solar masses, diameters 6 kpc to 100 kpc, and luminosities 10^8 to 10^{11} solar luminosities. The observed appearance of a spiral depends on the observer's point of view: Seen from above or below, a spiral looks basically round, but if viewed from the side, a spiral appears very flat, typically with an axial ratio $b/a \approx 0.1$. Making allowance for this, spirals still exhibit a far greater range of intrinsic shapes than do the ellipticals.

First, there is a fundamental distinction between spirals that show an axisymmetrical light distribution from center to edge (Hubble called these type S galaxies, but SA is probably preferred in a modern classification) and those whose centers are dominated by what appears to be a luminous bar across the center (barred spiral galaxies, type SB). The SA galaxies look like pinwheels with the spiral features curving symmetrically out of the nuclear region. The SB galaxies are typically two-armed spirals with the arms originating at the ends of the luminous bar crossing the central region. In making this distinction, Hubble actually identified the two extreme forms of spiral galaxies. About one-third of spirals show no evidence of a bar and are axisymmetric, about one-third have light patterns dominated by a bar, but the remaining third are intermediate in morphology, hence they are considered type SAB. Our own Milky Way has a bar in the center.

Spirals also show a wide range in the characteristics of the disk and its size in comparison to the central or nuclear bulge. Some galaxies have a bulge that is large relative to the disk (or, equivalently, a disk that is barely more extended than the nuclear bulge). In such galaxies, the spiral arms are barely visible, showing only a small contrast to the brightness of the rest of the disk. These spiral features also look thin and appear tightly wound about the center of the galaxy. Hubble labeled this subtype with the letter a, as in SAa and SBa (also termed early-type spirals for historical reasons). Other galaxies, labeled subtype b, show a less prominent bulge and a larger disk with more extensive spiral arms, more open and with a greater arm-interarm brightness contrast. Hubble's third subtype, c (late-type spirals), is represented by galaxies with hardly any bulge at all, with open, high-contrast spiral arms going right into the center of the galaxy. These three characteristics, the bulge-to-disk ratio, the openness of the winding of the spiral arms, and their brightness contrast tend to change with each other, although there are exceptions. In some modern versions of the Hubble classification are added types Sd (galaxies with no bulge, and spiral arms in a disk with barely enough symmetry to be called a spiral at all) and Sm (representing Magellanic-type irregular galaxies that have no particular symmetry; for example, a classification scheme considering the irregular galaxies to be an extension of the spiral types).

Although Hubble's classification again was based only on the optical appearance of galaxies, its utility lies in that the classification correlates with other galaxy properties. The Sa (the SAa and SBA galaxies together, making no distinction between the two) galaxies have little interstellar material, about 1 percent on average, and show a low rate of current stellar formation, correlating with the low brightness contrast of the spiral arms. Sb galaxies are more typically about 3 percent interstellar matter and have a greater rate of star formation, hence brighter spiral arms. Sc galaxies are even more gas rich, about 10 percent, and have even higher rates of star formation. That Sd galaxies are typically 20 percent interstellar material and Sm (=Im) galaxies are closer to 50 percent suggests a natural extension to the spiral types defined by Hubble.

Regardless of the type of spiral galaxy, in their disks it is the rotational motion of the stars in nearly circular orbits that produces the balance against gravity. The circular velocities are typically a few hundred kilometers per second.

Irregular (Ir) galaxies

Irregular galaxies (**Ir**) show little, if any, symmetry in their luminosity structure; their appearance really does appear irregular, and therefore they were defined by Hubble as a separate class of galaxy. In modern modifications of Hubble's classification system, some astronomers consider them to be a morphological extension of the spiral types of galaxy. Irregulars represent about 15 percent of all galaxies. These are mostly relatively low-mass systems, with 10^7 to 10^{10} solar masses or so, and contain the greatest fraction of interstellar material of any of the galaxies, up to 50 percent in some cases. Structurally, these are flat galaxies whose mass distributions are actually more symmetric than their light distributions. The high gas content is responsible for the greater rate of star formation. Where star formation does take place, there is a greater contrast in the surface brightness between the star-forming regions and the non–star-forming areas. These are also small galaxies in which the inward pull of gravity can be balanced by relatively low rotational velocities. However, this in turn means little in the way of differential rotation, and therefore, star-forming regions are not smeared into spiral arcs, unlike the more massive spirals. In other words, the basic difference between the spirals and the irregulars is mass; the spirals are the high-mass, gassy disk galaxies, and the irregulars are the low-mass disk galaxies. Differences in the history and present manner of conversion of interstellar mass into stars and the consequent optical appearance directly follow from differences in the circular motions that are needed to balance gravity.

S0 galaxies

A fourth type of galaxy, the **S0** ("ess-zero") is recognized as being distinct in appearance from both the spirals and ellipticals, though this type shares some characteristics of each. The S0 galaxies have smooth light distributions, like the ellipticals. On the other hand, they are definitely flat systems that are more like spirals containing both a halo population of stars (S0 galaxies show nuclear bulges) as well as a disk population of stars. Their rotational characteristics are like those of the faster rotating spirals and the surface brightness fades away toward the edge in the same manner as the spirals. As for other properties, these galaxies have intermediate sizes, masses, and luminosities; that is, no truly giant or truly dwarf S0 types are found. In Hubble's interpretation, these galaxies are composed only of stars, with no interstellar gas, and consequently no star formation-defining spiral arm regions. The S0 galaxy (and its barred counterpart, the SB0) were considered to be an "intermediate" or "transition" form of galaxy between the ellipticals and spirals. In the modern understanding of galaxies, this interpretation has been called into question, because it is now known that there exist apparently perfectly normal S0 galaxies that have significant fractions of their mass in the form of interstellar gas.

Hubble "tuning fork" diagram for classification

The purpose of any classification is not only to separate objects into distinct classes but also to seek an understanding of the relationships between the classes. Two aspects of the Hubble galaxy types are suggestive of a progressive relationship between the several types. The first is the distinction between pure stellar systems versus those with some content of interstellar material. Second, but related to the first, is a recognizable trend from "round" to "flat" galaxies. To visually portray the different types of galaxies in a simple manner, Hubble placed the round elliptical galaxies at the left and set the progressively flatter galaxies to the right, with the axisymmetric and barred spiral galaxies placed along two parallel paths. Arranged in this manner, the galaxies form what looks like a tuning fork on its side; that is, a "tuning fork" diagram (see Figure 14-2).

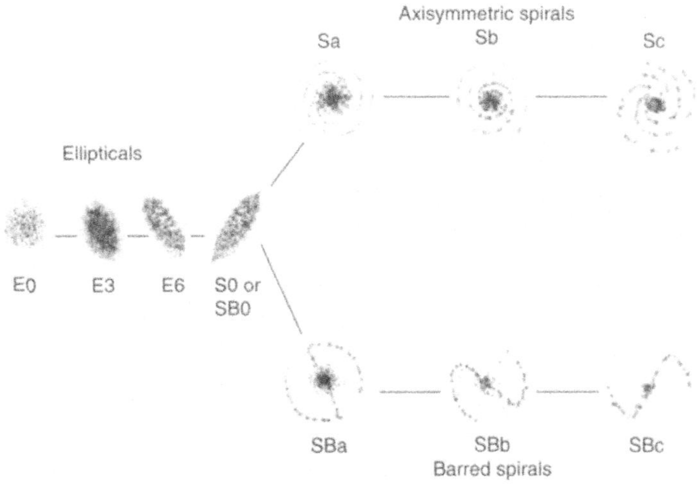

Figure 14-2

Hubble "tuning fork" diagram for galaxies.

As popularly portrayed, the Hubble diagram is misleading. Spiral galaxies are neither highly symmetric about the center (axisymmetric) or barred, but they show a full range of appearances from non-barred to barred. In other words, the area between the two prongs of the fork is actually filled with galaxies.

Originally, Hubble believed that his diagram mapped a sequence of evolutionary stages of galaxies, from an initial structure as an elliptical (hence the term early-type galaxy) followed by continued flattening into a final thin disk (late-type spiral galaxy). This perspective is incorrect because all galaxies show evidence of the same large age; every galaxy has a population of old stars. Even the irregulars have old stars whose faintness makes them hard to see under the light contributed by the bright young stars. If there are young galaxies, they are exceptions and are the result of special circumstances, such as collisions or mergers. Observational studies show large differences between the galaxy types in rotational properties (random motions are dominant in ellipticals but inconsequent in spirals), total masses (the largest ellipticals are more massive than the largest spirals; even among spirals, the most massive Sa galaxies are larger than the most mass Sc galaxies), mass distributions (pure "halo" for ellipticals, halo plus disk in different proportions for the spirals), and content ("old" stars and no interstellar material in ellipticals; old and young stars and a range of interstellar material in the spirals). Theoretical studies confirm that galaxies cannot change form in any simple manner from one type to another. Unlike stars that do evolve from Main Sequence to giants to white dwarfs, galaxies cannot evolve from ellipticals to spirals or vice-versa.

The galaxy types therefore appear to be fundamentally different for reasons other than dynamical evolution. A deciding factor appears to be the fraction of material in old stars versus old/young stars in disks (as indicated by the halo/bulge to disk size comparison), which would necessarily be related to the history of the conversion of interstellar material into stars in each galaxy type. These differences in the proportions of the halo and disk would suggest that galaxy types are due to differences at the time of formation, or slightly different primordial conditions leading to different final forms of galaxies. For example, it was suggested that an initial pregalactic cloud of gas with virtually no rotation would collapse into an elliptical galaxy, its gas going completely into stars during the collapse. If there were some rotation, then conservation of angular momentum would cause the gas to flatten into a rotating disk whose differential rotation would act to oppose star formation, hence a spiral galaxy with star formation continuing to the present time. The use of high speed computers to test these ideas numerically shows that it is impossible to form gas-free elliptical galaxies from an initial gravitational collapse. Star formation from the interstellar material is very inefficient, the remaining gas flattens to a disk, and only disklike gassy galaxies (not necessarily identical to the spiral galaxies of today) are formed. As elliptical galaxies do exist, then they must form in some manner from these original disk galaxies, a process that has come to be understood only by study of those galaxies that cannot be fit into the four types defined by Hubble.

Peculiar Galaxies

By definition, a **peculiar galaxy** is not a normal galaxy; for example, one that can be classified into one of the categories defined by Hubble. Astronomers have identified two types of peculiar galaxies, interacting galaxies and active galactic nuclei (AGN).

Interacting galaxies

The first type of peculiar galaxy are the **interacting galaxies,** or those galaxies whose structures have been altered by the gravitational attraction of neighboring galaxies.

There is no general classification possible for interacting galaxies simply because of the innumerable ways in which two galaxies can affect each other. A small galaxy will have little effect on a larger one (though it can stimulate a grand design two-armed spiral pattern), but the smaller galaxy may have its whole structure distorted or even be completely pulled apart by the tidal gravitational field of the larger galaxy (as is occurring for the Magellanic Clouds). Bridges and tails of luminous material may be produced. Totally disrupted, the stars and gas of a small galaxy may orbit a large galaxy perpendicular to the plane of its disk, forming a **polar ring galaxy.** These phenomena are short-lived by astronomical time scales, and ultimately the mass of the destroyed galaxy becomes part of the larger system, a case of galactic cannibalism (see Chapter 13).

A special case of a small galaxy dramatically but temporarily changing the appearance of a large galaxy occurs when the smaller galaxy's path is by chance nearly along the rotational axis of a large disk galaxy. As the small galaxy passes through the disk, its extra contribution to gravitation pulls in stars and interstellar material from the outer disk of the large galaxy. As the small galaxy departs, this material moves back outward. The momentary compression of the interstellar gas, however, triggers major star formation and the appearance of a **ring galaxy.**

Gravitational interaction that disturbs the mass distribution in a galaxy may also trigger major events of nearly simultaneous star formation over the whole disk of a galaxy. Such activity explains **starburst galaxies** such as M82. Many nearly simultaneous supernovae would greatly disrupt the remaining interstellar matter pattern, producing what looks like an exploding galaxy.

Mergers and cannibalism

More dramatic is the strong interaction that can occur when two large galaxies of similar masses interact. Even if the galaxies are originally in orbit about each other, energy exchange between orbital motion and the motions of the stars in each galaxy will cause the two to approach each other. This will result in the two galaxies ultimately merging (which will likely occur for the Milky Way and the Andromeda galaxies). If such a merger is relatively gentle and the galaxies are oriented properly with respect to each other, the final object will most likely resemble a spiral galaxy of some type. If the collision is more dramatic (head on; galaxies rotating in opposite directions; and/or the planes of rotation perpendicular to each other), little trace of the structures of the original colliding galaxies may remain. Energy exchange will randomize stellar orbits, but stellar collisions would be rare, because stars are so small compared to their average separations. The resulting distribution in space of the stars in the surviving single galaxy will not have an identifiable plane, nor will there be circular symmetry.

The effect on the interstellar gas when two galaxies merge would be more dramatic than the effect on the stars. Even with typical densities of one particle per cubic centimeter in each galaxy, the gas particles will collide, kinetic energy going into internal energy that then will be radiated away. Strong shock fronts will develop and, contrary to the normal inefficiency of conversion of interstellar material into stars, there can be complete conversion of the gas into stars. Within a few tens of millions of years of the merger, the bright blue stars will have evolved and disappeared, leaving behind a redder, stellar population of stars. Such an object would have all the characteristics of what Hubble classified as an elliptical galaxy. If, indeed, ellipticals are the result of mergers of disk galaxies, then observation of the population of galaxies early in the universe should show a smaller fraction of ellipticals than are present in today's universe. Studies with the Hubble Space Telescope have confirmed that this does appear to be the case.

Active Galactic Nuclei (AGN)

The second group of peculiar galaxies are those whose appearances and characteristics have been modified by internal processes, energetic events that appear to occur in the very nucleus of the galaxies. These are termed **active galactic nuclei objects,** or **AGNs** for short.

There is a vast range in the energetics of nuclear violence in galaxies. At the lower end of the energy spectrum is the phenomenon occurring in the nucleus of the Milky Way Galaxy. At higher energy levels are hot jets of material streaming out of galactic nuclei at nearly the speed of light (a so-called relativistic velocity as the rules of the theory of special relativity must be used to describe phenomena at these exceedingly high velocities). Jets are observed in visible light and in the radio region of the spectrum, for example, in the giant elliptical galaxy M87 in the Virgo Cluster of galaxies. The characteristics of the radiation identify it as **synchrotron radiation,** emitted by charged particles under the influence of a weak magnetic field. The appearance of a single bright jet is an illusion, however. When luminous material moves nearly at the speed of light, it preferentially emits its light in the direction of travel. The M87 jet is roughly pointed in the direction of Earth, but there is evidence for a faint counterjet on the opposite side of the galaxy nucleus. So there are actually two jets—a bipolar gas outflow, moving in opposite directions from a central source. The exact nature of this central source is unknown, but observation of normal stars very near the center of this galaxy shows large motions, which would imply (using Kepler's Third Law) an immense mass concentrated in a very small volume. The argument for a large mass is now believed to be evidence for the existence of a massive, nonstellar black hole that presumably formed at or near the time of the origin of the galaxy.

If the same phenomenon were to occur in the center of a spiral galaxy, however, the external appearance of the event would be very different. Unlike an elliptical galaxy, a spiral galaxy has significant interstellar material. Jets emanating from the center would immediately collide with the interstellar material, their energy stirring up the gas and heating it. Such spiral galaxies, identified by having exceptionally bright central regions (especially in the infrared), variable central luminosity, strong emission lines, and random gas motions up to thousands of kilometers per second, are known as **Seyfert galaxies.**

Given sufficient time, if there is no interstellar material to collide with or block them, jets can move outward to distances that are large with respect to the spatial distribution of the stars in the galaxy. The gas in the jets will cool and expand into volumes that may be several times the size of the central elliptical galaxy. Such gas still emits radio radiation and may be observed in large blobs (the radio astronomers call these lobes) on either side of a galaxy. If a **radio lobe galaxy** is drifting through space, the lobes may be trailed behind, forming butterfly-like structures with a size up to 1,000,000 pc in length, the largest individual objects known in the universe.

At the most extreme energies are the quasi-stellar objects (so named because they look star-like), shortened to **quasars,** or **QSOs.** The widely diverse nature of QSOs has led them to be identified under a number of terms including designation as **BL Lac objects** and **blazars,** although fundamentally these are believed to be the same type of object. They are characterized by high variability of their light output, with luminosities equivalent to that of a whole galaxy but coming from within a region of the size of the solar system. Careful photography reveals the faint outer luminosity surrounding the quasars showing that these actually are an occurrence in the nuclei of galaxies.

Quasars do not exist in the present-day universe, but were instead a phenomenon of the past, a fact revealed by their great velocities of recession, which places them at large cosmological distances. Given estimated quasar lifetimes of tens to hundreds of millions of years, statistical studies suggest that the number of ancient quasars is roughly equal to the number of present-day galaxies.

Most mechanisms for producing energy (gravitation, thermonuclear reactions, as in stars) require a great deal of mass, which in the case of quasars is prohibitive because of the constraint on the volume of space from which this energy must come. The most efficient energy source is the conversion of matter into energy via infall into a black hole. Any falling mass gains kinetic energy as it drops, but mass falling toward a small black hole, close to which the gravitational field becomes immense, gains a tremendous amount of kinetic energy—in fact an amount that is equal to the energy equivalence of the mass itself as given by the Einstein formula $E = mc^2$ (in contrast, the fusion of hydrogen to helium releases only 0.7 percent of the mass in the form of energy). In the right circumstance, this energy can be released as electromagnetic radiation. Thus quasars are thought to represent the era of formation of large central black holes in galaxies, from tens of millions to perhaps as much as a billion solar masses.

Such masses would affect only the very central conditions of a galaxy, and would have little effect on the structure and motions in the rest of the galaxies, thus they could be "hidden" in the centers of present-day galaxies, essentially invisible if no longer actively gaining new mass. The re-excitement of such a "central engine" by renewed mass infall (due to gravitational interaction with neighboring galaxies) could explain the lower energy phenomena seen in the Milky Way and other present-day galaxies (AGNs).

Clusters of Galaxies

Galaxies are like stars in that they tend not to be alone, but occur in small groups or larger clusters of galaxies. Clusters may be comprised of a few galaxies (for example, Stephan's Quartet) or massive systems, such as the Virgo Cluster with some 3,000 members and a total mass of 10^{15} solar masses. The smallest clusters have galaxies distributed irregularly through relatively small volumes of space, whereas the largest clusters have their many galaxies quite symmetrically distributed about the centers. Such regularity of form argues that motions within these clusters are in a balance against the self-gravity of the cluster members.

While clusters of galaxies were once considered the next larger hierarchical structure to be found in the universe, modern studies show that the actual distribution of galaxies in the universe is far more complicated. There are actually sheets (or walls) of galaxies surrounding vast voids that are nearly deficient of galaxies. Clusters and chains of galaxies and clusters of galaxies are found across space where such walls intersect. The actual distribution of visible galaxies in the universe can more properly be compared to the interior structure of a sponge.

The Local Group

The Local Group consists of about 30 galaxies distributed across about 1 Mpc of space. Immediately exterior to the Local Group, galaxies become sparse until the next small groups are encountered at distances of a few Mpc. Most of the mass of this small group is concentrated in the Milky Way Galaxy and its somewhat larger companion, the Andromeda Galaxy, best described as a binary galaxy system. Each of these large spirals is surrounded by a family of smaller dwarf galaxies (their coronas). About half a dozen small galaxies have a wider distribution and appear not to be specifically linked to either of the large spirals, although there is a group of a few galaxies (called the IC 342 group after the catalog number of the largest member) that models suggest may have been gravitationally slingshot away from the Local Group by the Andromeda Galaxy millions of years ago.

Giant clusters of galaxies

At one extreme, giant clusters of galaxies include systems like the Virgo Cluster, which contain some 3,000 members distributed loosely over several Mpc of space; most of these galaxies are spiral galaxies, although the largest galaxies in the cluster are two centrally located elliptical galaxies. At the other extreme are more compact systems like the Coma Cluster, which consists of thousands of galaxies, including about 300 large elliptical galaxies. The types of galaxies that are present are related to the environment in which they are found. Ellipticals are common where galaxies are closely packed together, but spirals predominate in loose, widely spread clusters. Such distribution of galaxy types supports the idea that ellipticals have formed over time from mergers of smaller disk galaxies.

Determination of mass in galaxies

For individual galaxies, measurement of optical rotation curves over the visible images of galaxies (this is a traditional method of study) can be done out to radii of maybe 20 kpc with typical masses on the order of 10^{11} solar masses deduced from the relationship $M(R) = V^2(R) R / G$. With the application of highly sensitive radio astronomy techniques in the 1970s, rotational velocities from neutral hydrogen gas could be detected to radii of 40–50 kpc. Galaxy rotation curves stay constant at large radii; doubling the distance outward for velocity measurement results in doubling the amount of mass associated with galaxies, but the radio observations also show this mass is not hydrogen gas.

Even larger volumes of space associated with individual galaxies may be measured for mass by use of binary galaxy studies. These studies must be done statistically because orbital periods are far too long for actual measurement of individual orbital sizes and periods that are necessary for direct application of Kepler's Third Law. Observation of a large number of galaxy pairs and judicious averaging of the measured separations between galaxies and velocities differences yield effective mass measurements out to an average of 80 kpc, suggesting even larger amounts of mass in each galaxy, $\sim 4 \times 10^{11}$ solar masses.

In the great clusters of galaxies, the **Virial Theorem** may be used. This principle relates gravitational effects (dependent upon mass and position of galaxies) to the kinetic energy of motions when those motions are random, not circular. Positions of galaxies can be measured in the plane of the sky and the Doppler effect gives the motion along the line of sight. The virial theorem then yields the one factor that cannot be measured directly, but must be inferred, that is the average mass for a typical galaxy. Virial Theorem studies suggest even larger masses of 8×10^{11} solar masses typically over radii of 150 kpc associated with each galaxy. In other words, these dynamical studies indicate that seven-eighths of the total mass of a typical galaxy may be outside the visible image of the galaxy, existing in some form that does not emit some form of detectable electromagnetic radiation. As additional studies confirmed these results, this was termed the missing mass problem, now known more appropriately as the **dark matter problem.**

And this indeed is a major problem facing modern astrophysics, for every condition of normal matter (density, temperature, and so on) can be observed by its emission of some wavelength of electromagnetic radiation. If something does not emit electromagnetic radiation, it almost certainly cannot be the familiar matter that makes up the stars and interstellar matter that otherwise are so well understood.

Dark matter

Neutrinos are among the factors that have been suggested to account for the dark matter associated with galaxies. Many neutrinos have been produced by thermonuclear reactions in stars and especially in the early era of the universe when the primordial helium was formed. If neutrinos have a tiny bit of mass, they are so numerous they could end up being the dominant mass of the universe. And they interact so weakly with everything else that they are essentially invisible. On the other hand, neutrinos probably are too energetic to allow gravitational clumping into galaxy-sized objects; they simply escape the cluster, and cannot provide the local source of gravity that is needed for clumping. Hence neutrinos are likely not a good explanation for dark matter.

Black holes may also account for the dark matter associated with galaxies. However, the black holes would not be those produced by the evolution of massive stars (there are too few of them), but rather unexpanded pieces of the early, high-density universe that were somehow left behind when the rest of the universe expanded. How these pieces were left over has yet to be explained.

There are also currently active observational searches for **MACHOs** (**Massive Compact Halo Objects**). These could be normal matter objects, so small (for example, compact) that their surface areas would produce so little electromagnetic radiation that they would be effectively invisible. So far, the preliminary results are significantly improving astronomers' understanding of the halo of the Milky Way Galaxy, but have not revealed the existence of a hitherto unknown type of object.

Theoretical studies, primarily aimed at understanding the formation of galaxies and replicating their distribution and motions throughout the universe, can be done without knowing the precise nature of the dark matter. Astronomers can define two generic types of dark matter: **cold dark matter,** which moves relatively slowly and is affected by gravitational clumping (like normal matter, it easily forms galaxies), and **hot dark matter,** which moves too quickly to respond to gravity (and therefore tends to slow the formation of galaxies and clusters of galaxies). Computation results at the present time suggest that the best comparison of theory with the real universe requires the presence of both cold dark matter and hot dark matter, but the ultimate understanding of this problem is yet to be settled.

The Origin and Evolution of Galaxies

With the discovery of the nature of galaxies, the first hypothesis developed to explain their existence was one of gravitational collapse in the primordial gas. As the forming galaxies grew smaller, the gas tended to fall into a flat plane, with fragmentation into stars occurring during both the collapse phase and continuing after formation of the final disk. The formation of a galaxy was completed when the mass distribution came into equilibrium between motions and gravity. The differentiation between the types of galaxies was thought to have been the result of initial conditions. If lots of angular momentum were present, a disk galaxy was produced. If initially there was little angular momentum, all matter became stars during the collapse phase, resulting in an elliptical galaxy.

Observational and theoretical work in more recent times has shown that galaxy formation is a much more complicated process. First, the efficiency of star formation is low. As a result, elliptical galaxies cannot be produced as was once thought; galaxy formation produces disk galaxies with significant interstellar material left over. Second, interactions between galaxies over the history of the universe can be significant. Galaxies do merge, and they cannibalize smaller companions. Violent interactions between disk galaxies appear to randomize motions and also to efficiently convert colliding interstellar gas into stars, leaving behind gas-free elliptical galaxies. Galaxies that may have grown in size, but avoided major disruptive encounters, appear to have evolved into the spectrum of spiral galaxies that exist today. Gentle encounters between two gassy disk galaxies are possible, and these encounters leave their fundamental stellar distributions unchanged but result in the gas being swept out, thus producing the relative rare, flat, gas-free galaxies known as S0s.

It is now hypothesized that the early era of galaxies was much more turbulent than today's universe. The process of producing equilibrium galaxies was associated with the growth of massive, nonstellar black holes in the nuclei. The liberation of tremendous energies during their formative stages is observed as quasars, but quasars died when galaxies achieved their equilibrium structures and ended mass infall into the centers. When new mass falls into the centers of galaxies, the central black hole phenomena can be re-ignited, explaining the active galactic nuclei of the present day.

CHAPTER 15
THE UNIVERSE

Observational Cosmology

Cosmology, the study of the whole universe, has been the endeavor of scholars since the earliest recognition that something existed beyond the limits of Earth. Humanity's concept of the cosmos, however, has changed dramatically over time. The earliest cosmologies dealt only with what is now known as the solar system. The recognition of the existence of distant stars culminated in the **Kapteyn Universe** concept, which implied that the universe was a flattened sphere of stars, some 17,000 pc (55,000 ly) in diameter and 4,000 pc (13,000 ly) thick, with the Sun near the center. Only in the early twentieth century was the nature of other galaxies established. Modern cosmological investigation presently is attempting to discern the history of the universe back to its very beginning. Astronomers using the Hubble Space Telescope and the new generation of large telescopes seek to discover the most distant galaxies that are observable; because of the finite speed of light, these objects are seen not as they are today, but as they were in the distant past. Earlier times in the history of the universe must be deduced theoretically.

Distances

Fundamental to understanding the universe is the distance scale to galaxies, which relies upon astronomical understanding of the stars that make up the galaxies. If a star of known type (that is, whose properties are identifiable with a specific kind of star in the Milky Way Galaxy) can be observed in another galaxy, then that star's apparent brightness compared to its known absolute luminosity will yield the distance to that galaxy. The most accurate measurements are derived from those stars, **standard candles,** whose absolute luminosities are well defined. Stars whose properties are calibrated by study of the Milky Way but that may be identified in other galaxies are also termed **primary distance indicators** and include RR Lyraes, Cepheid variables, novae, and planetary nebulae. The Cepheids are the most important. With the Hubble Space Telescope, astronomers can detect the longest period Cepheids in galaxies out to distances of about 20 Mpc.

At greater distances, even brighter or larger objects in galaxies must be found in order to infer distances. The properties of these **secondary distance indicators** are calibrated in galaxies whose distances have been obtained by use of the primary indicators. These galaxies include diameters of the largest HII regions, the brightest blue OB stars, the brightest red giant stars, and the brightnesses of supernovae Type I (the explosion of white dwarf stars whose mass has exceeded the Chandrasekhar Limit). Of these, the supernovae appear to produce the most reliable results for cosmological investigation.

With astronomers obtaining distances to a large sample of galaxies by use of the secondary indicators, they can do calibration of the properties of the brightest and largest galaxies. These **tertiary distance indicators** include the largest of the giant elliptical galaxies in clusters of galaxies (first ranked cluster ellipticals), as well as giant Sc spiral galaxies.

Other techniques used to infer distance depend upon global properties of galaxies. In spiral galaxies, the maximum rotational velocity is correlated with the mass of the galaxy. Similarly, the absolute luminosity is related to mass. The **Tully-Fisher Relationship** uses the maximum rotational velocity (which can be obtained by application of the Doppler effect) to deduce the absolute magnitude of the galaxy; comparison of the absolute magnitude with the object's apparent magnitude then yields the distance. In a similar fashion, the **Faber-Jackson Relationship** yields the absolute magnitude of an elliptical galaxy based upon a spectroscopic determination of its internal random motions.

The Hubble Relationship

The basis of modern cosmology was established with the recognition in 1929 that absorption features in the spectra of more distant galaxies are shifted progressively toward the red end of the spectrum. That is, the more distant the galaxy, the greater the wavelength displacement. The most apparent explanation of this **redshift** is a result of a Doppler shift (see Chapter 2) due to a motion away from the observer, with more distant galaxies moving away faster. Distances and line-of-sight velocities are simply related as

Doppler velocity = constant x distance

or expressed algebraically

$V = HD$

This relationship is known as the **Hubble Law** and the constant of proportionality H is the **Hubble Constant.** (See Figure 15-1.)

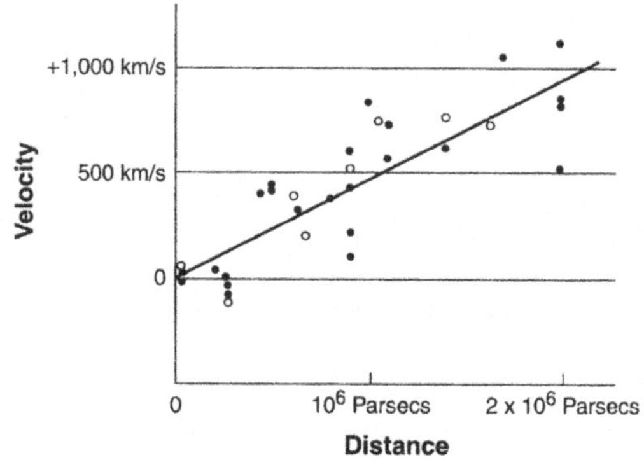

Figure 15-1

The Hubble Relationship and Law (Hubble 1929).

The value of the Hubble Constant represents not only the present rate of expansion of the universe, but also a distance scale for the universe and a measure of the age of the universe. The distance to a galaxy or any other object at a **cosmological distance** is given by the Hubble Law expressed in the form

$$D = V/H$$

and requires only a spectrum for which a Doppler shift can be measured. With his understanding of the Cepheid Period-Luminosity Relationship, Hubble in 1929 obtained a value of H = 560 km/s per Mpc. With various refinements and corrections in the data, most recent studies suggest a much smaller value of 50 km/s per Mpc $<$ H $<$ 100 km/s per Mpc with the most likely value about H = 60 km/s per Mpc, that is, the expansion velocities increase by 60 km/s for each increase of 1 Mpc (3,260,000 ly) in distance. In terms of distances, a change in the Hubble Constant represents a change in astronomers' understanding of the distance scale or size of the universe. A galaxy with a recession velocity of 2,000 km/s, for example, is considered to be at a distance of 20 Mpc if H = 100 km/s per Mpc, but at twice this distance, 40 Mpc, if H = 50 km/s per Mpc.

An estimate of the age of the universe may be made by calculating how long a galaxy has taken to achieve its present distance away from the Milky Way. Travel time T is simply distance D traveled divided by the travel velocity V, but as the Hubble Law shows the relationship V = H D

$$T = D/V = D/(H\,D) = 1/H$$

The travel time is independent of present distance! More distant objects are more distant because they have been moving away at a faster pace. Because the self-gravity of the universe should be slowing its expansion, in standard cosmological theory this "age" for the universe (known as the **Hubble Time** $T_H = 1/H$) is an *upper limit* to the time that has elapsed since every position in the universe was superimposed upon every other position, for example, since it had zero extent or infinite density. This is a problem because physicists don't like working with infinities; an infinite quantity is not physical, or real! The initial infinity at which point the cosmological clock is set (the "origin" of the universe) is the result of theorists' limited understanding of physical law. Most theoretical models for the evolution of the universe suggest a more realistic "age" $T_{actual} = \frac{2}{3} T_H$; thus, if $H = 100$ km/s per Mpc, then $T_{actual} = 6.7$ billion years, but if $H = 50$ km/s per Mpc, $T_{actual} = 13.3$ billion years. The expansion of the universe and its finite time of existence in its present form are the major factors that a theory of cosmology must deal with. It is interesting, however, that these implied ages appear to conflict with the older ages suggested by stellar evolution theory for the oldest stars in the Milky Way Galaxy. Clearly, one or the other (or both!) of the theories is not completely correct. Only additional observation and refinements in theoretical understanding will eliminate this discrepancy.

Olbers' Paradox

If the universe were infinite and filled uniformly with stars, then in every direction we would see a star and hence every point in the sky should be bright. This is **Olbers' Paradox,** an apparent contradiction between the darkness of the sky and the theoretical expectation that the sky should be bright. Like every other so-called scientific paradox, Olbers' Paradox reflects not a contradiction in the nature of reality, but an error in interpretation of natural phenomena. That the sky is dark must reflect the fact that the universe is not infinite, or the universe has a finite age (the greatest lookout distance is given by the speed of light times the age of the universe, or $c \times T_{actual}$). Olbers' Paradox is thus consistent with the inferred evolution of the universe as implied by the Hubble Law.

3 K cosmic background radiation

An expansion of the universe implies a high density and temperature in an early era, before a time of about 100,000 years. Earlier than this, at any time the spectrum of black body radiation was characteristic of the high temperature, but this radiation cannot be observed today. As the temperature of the universe cooled, the spectrum of the photons also continually changed, at any instant the maximum in the spectrum of radiation (Wien's Law) simply indicating the current temperature. All spectral record of prior higher temperatures was erased. Only when the universe became transparent (radiation no longer interacting strongly with matter), did the spectrum become permanent with a peak in the gamma-ray part of the spectrum. Today this radiation is observed as if coming from a large distance and hence subjected (the Hubble Law) to a large redshift. The shape of the spectrum is preserved, but now this **cosmic background radiation** is observable as black body radiation indicative of a much lower temperature, with its peak radiation in the radio or microwave region of the spectrum (hence the alternate name **cosmic microwave radiation**).

This **cosmic relic radiation** was discovered in 1965 when scientists examined interference occurring with the first satellite communication systems. The launch of the **COBE (Cosmic Background Explorer) satellite** confirmed the shape of the black body spectrum at a temperature of 2.728 ±± 0.004 K. COBE showed this radiation is too uniform to be coming from myriads of sources, each too faint to be observed individually; thus, it must be coming from the early universe when all material was spread out before gravitational collapse into galaxies. Superimposed on its uniformity, there is a **dipole effect** in the temperature, a small, smooth change from one direction across the sky to the opposite direction, the result of the motion of the Local Group of galaxies with respect to the **cosmic black body radiation**. There also are tiny variations, 1 part in 100,000, that show a lack of complete density uniformity in the material that emitted the radiation. The slightly denser regions subsequently collapsed under their self-gravitation to become galaxies, clusters of galaxies, and so on.

The Big Bang Theory

What has become known as the **Big Bang theory** originally was an attempt by George Gamow and his coworkers to explain the chemical elements in the universe. In this, the theory was incorrect because elements actually are synthesized in the interiors of stars, but the theory is still successful in explaining many other observed cosmological phenomena. Using the same physical principles for understanding stars, the theory does account for the evolution of the universe after a time of about 30 seconds. Those aspects that the Big Bang theory was developed to address are Olbers' Paradox, the Hubble Relation, the 3 K black body radiation and its present ratio of 10^9 photons for each nucleon, the apparent large-scale uniformity and homogeneity of the universe, the primordial helium-to-hydrogen ratio (even the oldest stars are about 25 percent helium, thus helium must have a prestellar origin), and the existence of clusters of galaxies and individual galaxies (that is, the small-scale variations in the mass distribution of today's universe).

Cosmological Principle

Two explicit assumptions are made in the Big Bang cosmological model. The first is that the observed shift of features in galaxy spectra to redder wavelengths at greater distances is really due to a motion away from us and not to some other cosmological effect. This is equivalent to saying the redshifts are Doppler shifts and the universe is expanding. The second assumption is a basic principle that the universe looks the same from all observing points. This **Cosmological Principle** is equivalent to saying the universe is homogeneous (the same everywhere) and isotropic (the same in all directions). This is the ultimate **Copernican Principle** that the Earth, Sun, and Milky Way Galaxy are not in a special place in the universe.

According to the Big Bang Cosmology, the universe "originated" at infinite temperature and density (not necessarily true, because the conventional rules of physics do not apply to the exceedingly high temperatures and densities at a time before 30 seconds, which was in a state that scientists are only now beginning to understand). Coming out of this early unknown era, the universe was expanding with both temperature and density decreasing. Initially the radiation density exceeded matter density (energy and mass have an equivalency given by $E = mc^2$), thus the physics of radiation governed the expansion.

For matter, the density relationship with respect to any measure of the size of the universe r is straightforward. Volume increases as $\text{length}^3 = r^3$. A fixed mass within an expanding volume thus has a density ρ = mass/volume, hence proportional to $1/r^3$. For electromagnetic radiation, the density of a fixed number of photons in a given volume changes in the same way that mass changes, or photon number density is proportional to $1/r^3$. But a second factor must be introduced. The energy E of each photon depends inversely on its wavelength λ. As the universe expands, the wavelengths increase also, $\lambda \propto r$; hence the energy of each photon actually decreases as $E \propto 1/r$ (this is a consequence of the Hubble Law: a photon moves at the speed of light, hence any photon is observed as having come from a distance and is subjected to a redshift). The evolution of the energy density therefore requires both factors; energy density $\rho \approx (1/r^3)(1/r) = 1/r^4$, so it decreases faster than mass density with its $1/r^3$ dependence. At some time in the history of the universe, the density of the radiation dropped below the density of the real mass (see Figure 15-2a). When this occurred, the gravitation of the real mass began to dominate over the gravitation of the radiation and the Universe became matter dominated.

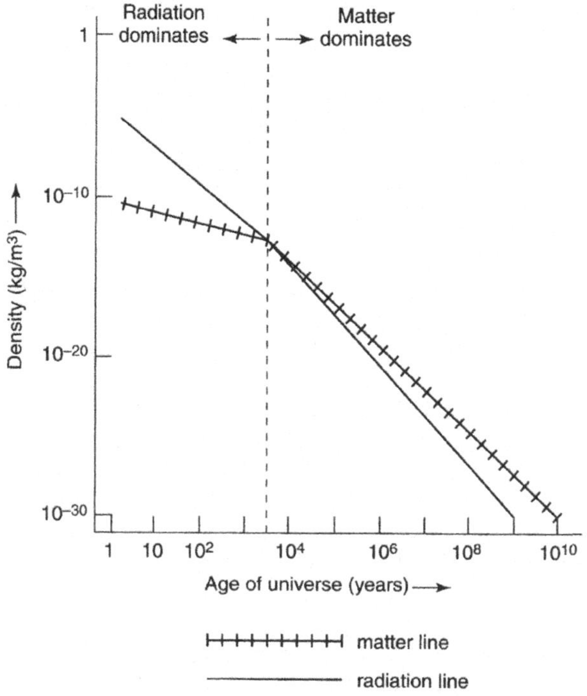

Figure 15-2a

Density of the evolving unverse.

At extremely high temperatures, normal matter cannot exist because photons are so energetic, the protons are destroyed in interactions with photons. Thus matter came into existence only by a time of about $t \approx 1$ minute when the temperature dropped below $T \approx 10^9$ K and the average energy of photons was less than what is necessary to break apart protons. Matter began in its simplest form, protons or hydrogen nuclei. As the temperature continued to drop, nuclear reactions occurred, converting protons first into deuterium and subsequently into the two forms of helium nuclei by the same reactions that now occur in stellar interiors:

$^{1}H + ^{1}H \rightarrow ^{2}H + e^{+} + \nu$

$^{2}H + ^{1}H \rightarrow ^{3}He$

$^{3}He + ^{3}He \rightarrow ^{4}He + ^{1}H + ^{1}H$

Also, a tiny amount of lithium was produced in the reaction

$^{4}He + ^{3}He \rightarrow ^{7}Li + e^{+} + \nu$

Heavier elements were not produced because by the time a significant abundance of helium was produced, the temperatures and densities had dropped too low for the triple-alpha reaction to occur. In fact, by $t \approx 30$ minutes, the temperature was too low for any nuclear reactions to continue. By this time, approximately 25 percent of the mass had been converted to helium and 75 percent remained as hydrogen.257

At high temperatures, matter remained ionized, allowing continual interaction between radiation and matter. As a consequence, their temperatures evolved identically. At a time of about 100,000 years, however, when the temperature dropped to $T \approx 10{,}000$ K, recombination occurred. Positively charged nuclei combined with the negatively charged electrons to form neutral atoms that interact poorly with photons. The universe effectively became transparent, and matter and photons no longer strongly interacted (see Figure 15-2b). The two **decoupled,** each subsequently cooling in its own way as the expansion continued. The cosmic black body radiation, about 1 billion photons of light for every nuclear particle, is left over from this **era of decoupling.**

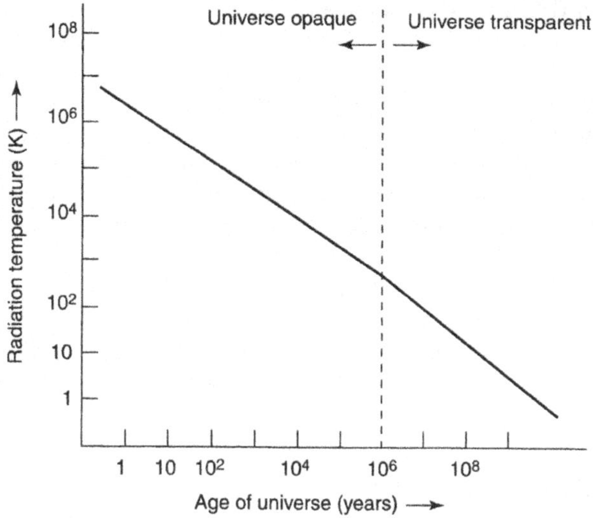

Figure 15-2b

Temperature of the evolving universe

By an age of 100 million years to 1 billion years, matter began to clump under its self-gravitation to form galaxies and clusters of galaxies, and within the galaxies, stars and clusters of stars began to form. These early galaxies were not like the galaxies of today. Hubble Space Telescope observations show them to have been gassy disk galaxies, but not as regularly structured as true spiral galaxies. As the universe continued to age, galaxies regularized their structures to become the spirals of today. Some merged to form ellipticals. Some galaxies, if not all, underwent spectacular nuclear region events, which we now observe as the distant quasars.

In the Big Bang theory, the present-day homogeneity of the universe is considered to be the result of the homogeneity of the initial material out of which the universe evolved; but this is now known to be a serious problem. For one region of the universe to be just like another (in terms of all physically measurable properties, as well as the very nature of the laws of physics), the two must have been able to share or mix every physical factor (for example, energy). Physicists express this in terms of **communication** (sharing of information) between the two, but the only means of communication between any two regions is one receiving electromagnetic radiation from the other and vice-versa; commication is limited by the speed of light. Throughout the whole history of the universe, regions that today are on opposite sides of the sky have always been farther apart than the communication distance at any era, which is given by the speed of light times the time elapsed since the origin of the universe. In the language of physicists, there is no **causal** reason for every region of the observable universe to have similar physical properties.

Closed and open universes

Within the context of a Big Bang theory there are three types of cosmologies that are differentiated on the basis of dynamics, density, and geometry, all of which are interrelated. An analogy may be made in the launch of a satellite from Earth. If the initial velocity is too small, the satellite's motion will be reversed by gravitational attraction between Earth and satellite and it will fall back to Earth. If given just enough initial velocity, the spacecraft will go into an orbit of fixed radius. Or if given a velocity larger than the escape velocity, then the satellite will move outward forever. For the real universe with a rate of expansion as observed (Hubble Constant) there are three possibilities. First, a low-density universe (hence low self-gravity) will expand forever, at an ever slowing rate. As mass has a relatively weak effect on the expansion rate, the age of such a universe is greater than two-thirds of the Hubble Time T_H. Second, a universe with just the right self-gravity, for example a **critical mass universe,** will have its expansion slowed to zero after an infinite amount of time; such a universe has a present age of $(2/3)T_H$. In this case, the density must be the critical density given by

$$\rho = 3H_0^2/8\pi G \approx 5 \times 10^{-30} \text{ g/cm3}$$

where H_0 is the Hubble constant measured in the present day universe (due to the gravitational deceleration, its value does change over time). In a higher-density universe, the current expansion at a time of less than $(2/3)T_H$ ultimately is reversed and the universe collapses back onto itself in the big crunch.

Each of these three possibilities, via the tenets of Einstein's theory of general relativity, are related to the geometry of space. (General relativity is an alternative description of gravitational phenomena, in which changes in motions are the result of geometry rather than the existence of a real force. For the solar system, general relativity states that a central mass, the Sun, produces a bowl-shaped geometry. A planet moves around this "bowl" in the same manner that a marble prescribes a circular path within an actual curved bowl. For mass distributed uniformly over vast volumes of space, there will be a similar effect on the geometry of that space.) A low-density universe corresponds to a **negatively curved** universe that has **infinite** extent, hence is considered **open.** It is difficult to conceptualize a curved geometry in three dimensions, hence two-dimensional analogs are useful. A negatively curved geometry in two dimensions is a saddle shape, curving upwards in one dimension, but at right angles curving downward. The geometry of a critical mass universe is **flat** and **infinite** in extent. Like a two-dimensional flat plane, such a universe extends without bound in all directions, hence also is **open.** A high-density universe is **positively curved,** with a geometry that is **finite** in extent, thus considered to be **closed.** In two dimensions, a spherical surface is a positively curved, closed, finite surface.

Will the universe expand forever?

In principle, observation should allow determination of which model corresponds to the real universe. One observational test is based upon deducing the geometry of the universe, say by number counts of some type of astronomical object whose properties have not changed over time. As a function of distance, in a flat universe, the number of objects should increase in proportion to the volume of space sample, or as $N(r) \propto r^3$, with each increase of a factor of 2 in distance producing an increase in the number of objects by $2^3 = 8$ times. In a positively curved universe, the number increases at a lesser rate, but in a negatively curved universe, the number increases more rapidly.

Alternatively, because the strength of gravity slowing the expansion of the universe is a direct consequence of the mass density, determination of the rate of **deceleration** constitutes a second potential test. Greater mass means more deceleration, thus a past expansion is much more rapid than at present. This should be detectable in measurement of Doppler velocities of very distant, young galaxies, in which case the Hubble Law will deviate from being a straight line. A lesser mass density in the universe means less deceleration, and the critical case universe has an intermediate deceleration.

Differing rates of expansion in the past also yields a direct relationship to the ratio of helium-to-hydrogen in the universe. An initially rapidly expanding universe (high-density universe) has a shorter time era for nucleosynthesis, thus there would be less helium in the present day universe. A low-density universe expands more slowly during the helium-forming era and would show more helium. A critical case universe has an intermediate helium abundance. Deuterium and lithium abundances also are affected.

The fourth test is to measure directly the mass density of the universe. In essence, astronomers select a large volume of space and compute the sum of the masses of all the objects found in that volume. At best, individual galaxies appear to account for no more than about 2 percent of the critical mass density suggesting an open, forever expanding universe; but the unknown nature of the dark matter makes this conclusion suspect. The other tests suggest a universe that is flat or open, but these tests are also fraught with observational difficulties and technical problems of interpretation, thus none really produces a decisive conclusion.

Is the Big Bang theory correct?
Recent observations of Type I supernovae in distant galaxies suggest that, contrary to a basic assumption of the Big Bang cosmological theory, the expansion may actually be accelerating, not slowing. Scientists always worry that a single suggestion in major conflict with accepted theory may itself be in error. One always wishes confirmation, and in 1999 a second group of astronomers was able to provide confirmation that the expansion is indeed accelerating. How this will force changes in cosmological theory is as yet unclear. The Big Bang theory has been very successful in describing many aspects of the history of the universe, but now we know that it is incomplete. Every astronomy textbook will need some major revisions in the next few years.

Beyond the Big Bang Theory

Although the general outline of the classical Big Bang cosmology has served well to provide an understanding of both the present nature of the universe and a large part of its past history (after a time of about 30 seconds), there are several matters that this theory currently cannot explain. One of these has already been mentioned in the discussion of the Big Bang cosmology, the **communication problem.** The large-scale uniformity of the properties of the universe requires that every region of the observable universe must once have been able to share information with every other region, a possibility ruled out by the finite speed of light and the nature of expansion in a Big Bang universe.

The existence of galaxies is actually also a problem. In the Big Bang theory, density fluctuations in the early universe that left their mark on the temperature fluctuations (1 part in 10^5) of the cosmic background radiation grew into the galaxies of today. But why did these density fluctuations actually exist at the time of decoupling? For the average density at that time, the statistical laws of variability, that is, random chance, require an exceedingly uniform universe, much smoother than observed! Some physical effect stemming from the even earlier universe must be responsible for beginning the rearrangement of matter from an earlier homogeneous density state to the weakly nonuniform state at the time of decoupling.

The very existence of normal matter represents a third problem. In the physics of the present day universe, there is a **symmetry** in the relationship between matter and energy (in the form of electromagnetic radiation). Nature, on the one hand, can create matter (and antimatter) in the reaction

high-energy photon ↔ matter particle + antimatter particle

and destroy both forms of matter through the reaction

matter particle + antimatter particle → high-energy photons

The two sides of each equation represent different aspects of what is essentially identical, and both reactions can be summarized in a single expression where the double-ended arrow indicates that the reaction is permitted to go in both directions:

high-energy photon ↔ matter particle + antimatter particle

The reaction can go back and forth any number of times, and after an even number of reactions (no matter how large), the physical situation is exactly where it started: Nothing has been changed, lost, or gained. Thus there should be no excess of one type of matter over the other, unless during an early epoch in the history of the universe the physics of the electromagnetic radiation-matter interaction was different. If the physical rules were different, then

photons → matter + antimatter → photons + matter

leaving behind in the present universe about one nuclear particle for every 10^9 photons.

Related to this is the question of the **dark matter,** or the invisible matter whose existence is postulated by astrophysicists to account for the large amount of observed gravitation that cannot be accounted for by visible matter. The dynamics of normal galaxies suggest that perhaps only 10 percent or less of the gravitating matter in the universe is observable with visible light or some other form of electromagnetic radiation that can be detected on Earth and from which the state of the material that emitted the radiation can be deduced. As every form of known matter, regardless of its temperature of other physical conditions, emits some form of this radiation, this matter must exist in some form not described by the physics of today's universe.

To all the other aspects of the universe scientists wish to understand would be the question of why there exist four distinct forces of nature. Gravity is the weakest of the four forces. Electromagnetism is some 10^{40} times stronger. The other two forces act at the nuclear level. The weak nuclear force is involved in electron reactions (such as $^1H + {}^1H \rightarrow {}^2H + e^+ + \nu$), and the strong nuclear force holds protons and neutrons together in the atomic nuclei.

A final problem is that the Big Bang cosmology alone is not able to address why the geometry of the universe is so close to being flat. The Big Bang cosmology allows for a variety of geometries, but makes no specification as to what the geometry should be. Observation suggests the geometry is very close to being flat, but this is a difficult result to understand. If the initial universe began ever so slightly different from being flat, then over its evolution until today the curvature should have become enhanced. In other words, some unknown cause very early in the history of the universe appears to have forced a flat geometry.

Grand Unified Theories (GUTs)

The apparent resolution to understanding the origin of these six additional aspects of the universe has come not from refinement of cosmological theory, but from theory aimed at understanding the interrelation between the four forces of nature and their further relation to the existence of the many types of particles that physicists have produced in high-energy particle accelerators (over 300 so-called elementary particles are now known). Each force appears to have an association with a particle that transmits that force: The electromagnetic force is carried by the photon, the weak force by the Z particle, the strong force through gluons. No one knows if gravity has an associated particle or not, but quantum theory predicts that the graviton does indeed exist.

Einstein tried (and failed) to unify gravity and electromagnetism. Modern theorists have succeeded in a theoretical unification of the electromagnetic force and weak force (theory of the **electroweak force**). In turn, various theoretical schemes (**Grand Unified Theories** or **GUTs**) to unite the electroweak force and strong force (into a **superforce**) are being investigated at the present time. Ultimately, the theoretical goal is to unite gravity and a Grand Unified Theory into a single theoretical formulism, a **theory of everything,** in which there would be a single unified force (for example, Quantum Gravity or Supergravity). Each stage of unification, however, occurs at successively higher energies and therein lies the cosmological connection — the early universe was a high temperature, high energy density situation at which time existed vast quantities of the exotic particles associated with each of these unifications.

From these recent theoretical developments, an outline of the very earliest history of the universe may be deduced. The universe began with a single (unified) force in existence, but the physics of this era before a time of 10^{-43} seconds will be known only when the final unification of gravity into the theory has been achieved. Before 10^{-43} seconds, the so-called **Planck time,** is an unknown era for which existing gravitational theory (general relativity) and Grand Unified Theories are in conflict. After this time, however, the expanding universe evolved monotonically to lower temperatures. As temperatures and energies dropped, the several forces became distinguishable in their behavior:

single unified force $\rightarrow$ superforce + gravity

superforce $\rightarrow$ electroweak force + strong force

electroweak force $\rightarrow$ electromagnetic force + weak force

This is a **symmetry breaking** in the sense that in the present universe, the opposite reactions, a recombination of these forces into, a single force, won't occur.

The Inflationary Universe. A major aspect of applying Grand Unified Theories to the early history is the recognition that the universe did not always expand at a rate that can be determined from observations of the present day universe. At an epoch of 10^{-35} seconds after the initial infinite density, it is theorized that there occurred a surge in the expansion, an **inflation** by perhaps 10^{30} times. In an instant, everything within the present-day observable universe (a diameter of about 9 billion parsecs or 30 billion light-years) went from approximately the size of a proton to the size of a grapefruit. Why? Because in the GUTs, the description of what we think of as space requires additional factors than things like familiar length, density, and so forth; more importantly as the universe evolved, these factors changed with the accompanying release of immense energy. In the jargon of physicists, one talks about there being a "structure" to the **vacuum** (this use of the word is very different from the normal usage of meaning "completely empty space"). As the universe expanded and the temperature dropped, the vacuum underwent a **phase change** from one state of existence to another. This change is analogous to the phase transition of water from gaseous steam to liquid. Liquid water is a lower-energy phase, and the energy released by water condensing from steam to liquid can produce work in a steam engine. In a similar manner, as the vacuum went from a high-energy to a low-energy phase, the energy released drove a momentary inflation in the size of the universe, followed by the much slower rate of expansion that continues today. This phase transition was responsible for the separation of the strong force from the electroweak force; in the higher-energy, preinflation state, these two forces were linked into a single force. In the lower-energy, postinflation state, the two forces are no longer identical and could be distinguished from each other.

There is a further significant consequence of the inflation that is important in understanding the present universe. Nearby regions that were in communication with each other before the inflationary expansion (the communication distance is the speed of light times the age of the universe), and that therefore had the same physical properties of energy density, temperature, and so on, ended up at a later time, after the rapid expansion, much further apart than estimated on the basis of using only the present expansion rate. Because these regions evolved over time, the laws of physics starting with their original similar conditions produced the present day similar conditions. This explains why regions now widely separated in opposite directions in our sky have the same properties even though these regions are no longer in communication (distance apart now being greater than the speed of light times the present age of the universe).

A second and more consequential result is present: The GUTs do allow a symmetry breaking in the interaction between matter and photons, allowing an excess of normal matter (proton, neutrons, and electrons—the material that makes up matter as we know it) to be present after the universe cooled to its present state. However, this is only part of the existence of gravitating material in the universe. GUTs force a major inflation in the universe. No matter how curved the early universe was, this inflation in size forces the universe to have a flat geometry. (By analogy, a basketball has a surface that is obviously curved, but if suddenly increased in size by 10^{30} times, making it about 1,000 times larger than the present visible universe, then any local area of the surface would appear very flat). A flat geometry means that the true density of the universe must be equal to the critical density that divides universes between those that will expand forever and those that will collapse back into themselves. Dynamical studies of galaxies and clusters of galaxies have been suggesting that 90 percent of the gravitating material of the universe is not visible, but all their matter, visible plus dark, if spread uniformly over the volume of the universe, yields only ~10 percent of the critical density. GUTs demand a density equal to the critical density, thus it is not 90 percent of the mass of the universe that is invisible, but 99 percent! (See Figure 15-3.)

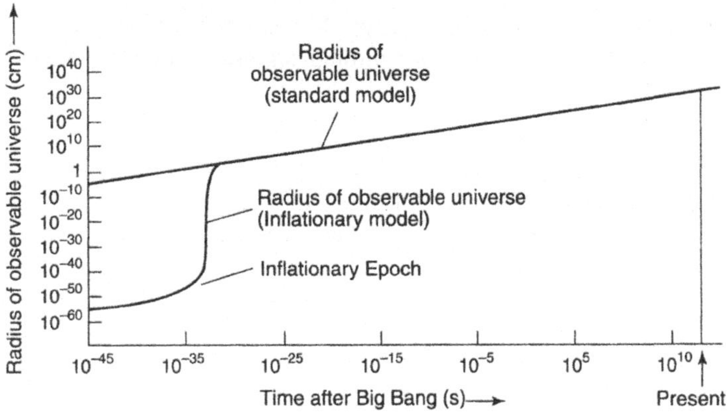

Figure 15-3

Evolution of the universe including the inflationary era.

Dark Matter. GUTs predict on the one hand far more dark matter in the universe than implied by studies of galaxies. But on the other hand, GUTs also predict the existence of many particles other than the material (protons, neutrons, electrons, photons) that make up the visible universe. An abundance of possibilities exist for the dark matter, depending upon which version of Grand Unified Theory you consider. Sophisticated physical experiments are being designed and put into operation to attempt to test for the existence of these possibilities, both to eliminate incorrect versions of GUTs as well as to identify the true nature of the dark matter. Some dark matter possibilities are WIMPs (**Weakly Interacting Massive Particles**), **axions** (lightweight particle types that again interact poorly with everything else), **strings** (features in the structure of space that are analogous to the boundaries between different crystals in a solid material), **magnetic monopoles** (in essence, incredibly tiny pieces of the early universe, with the conditions of temperature, energy, and the physical laws of the preinflation universe preserved behind a shell of exotic particles), and **shadow matter** (a second form of matter that has evolved independently of normal matter, whose presence is detectable only through its gravity). Which, if any, of these ideas are correct will be determined only through significant research effort.

Cosmological Constant

One additional factor may influence cosmological evolution. The mathematical equations describing the evolution of the universe allow for a **cosmological constant,** a factor originally introduced by Einstein. This factor would act as a repulsive force working against gravity. The evolution of the universe at any era would thus depend on which factor is stronger. It also is interpreted as an energy density of the vacuum, which would exist even if there were no matter and no electromagnetic radiation in the universe, hence another contributor to the dark matter. Most theory considers the cosmological constant to be zero, but its true value is yet to be determined. Ironically, Einstein introduced the cosmological constant erroneously; because he thought the universe was static and constant in size, he used the cosmological constant as a force to oppose gravity. Without it, he predicted the universe would collapse. However, a few years later it was discovered that the universe was expanding, and he realized the constant wasn't needed. He called it the biggest blunder of his life! The recent findings using Type I supernovae that the universe may be accelerating its expansion has reawakened interest in the cosmological constant. Future research and further observations will help shed light on this old problem.

CHAPTER 16
LIFE IN THE UNIVERSE

Historical Background

The consideration of whether humankind is alone in the universe or shares it with other creatures has long been part of intellectual inquiry into the nature of the universe. The early Greek philosophers, for instance, speculated about the existence of other inhabited planets. With the recognition that the other planets of the solar system are bodies like Earth came both scientific and popular speculation about life on other planets. Much of the modern effort of exploration in the solar system has been driven implicitly, if not explicitly, by this search for extraterrestrial life. Those who investigate the possibility of extraterrestrial life and the effects of extraterrestrial environments on living organisms from Earth are called **exobiologists.**

In the late 1800s, the Italian astronomer Schiappareli thought he saw linear features crossing the surface of the planet Mars. (These features are actually just the attempt of the eye and brain to find a pattern in random features seen at the edge of detectability). His "channels" became, in English translation, "canals" and became the inspiration for Percival Lowell to not only found his own observatory for their study, but to also widely lecture to audiences about intelligent life on a dying Mars. Lowell's viewpoint was the inspiration behind H. G. Wells's fictional novel *War of the Worlds*, which was dramatized in the infamous Orson Wells radio broadcast of October 30, 1938.

The recurrent fascination with Mars was a major factor behind the 1976 NASA Viking mission, which landed a spacecraft on the planet. Part of its task was to test Martian soil for evidence of the chemicals produced by living organisms, an experiment that produced a negative result. With Viking removing doubt about the possibility of major life forms on Mars and Venus proven to be too hot to be habitable, scientific speculation turned to the outer planet,

Jupiter, and its moons, as well as the cloud-enshrouded moon Titan, which orbits Saturn. The outer atmosphere of Jupiter is now known to be too turbulent to support a stable environment for life. Additionally, Jupiter's moons are probably too ice-enshrouded for life to be found in the oceans lying beneath the ice, and Titan has been shown to be too cold to support life. A recent (1996) report of the discovery of microscopic bacterial fossils in a meteorite that originated on Mars caused great excitement, but continued study suggests that the structures that were found are likely not organic in nature. Nevertheless, the announcement in early 2000 that high resolution photography of small areas of the Martian surface reveals extensive patterns of erosion features, apparently identical with drainage features on Earth due to water, has rekindled interest in the planet. If water does exist somewhere on the planet, then maybe the evidence for living organisms also is present.

Even if life does not exist elsewhere in the solar system, there are billions of stars and potential solar systems that could provide hospitable environments. With scientific proof yet to be found, nevertheless the most popular modern genre of written fiction, television, and movie entertainment is science fiction, with its worlds (*Star Trek, Star Wars,* and so on) populated by a vast variety of other creatures.

Scientific Considerations

Although no direct evidence has been uncovered to show that life exists elsewhere, scientists present reasonable arguments to postulate what may be. These arguments, for the most part, are based on the **Principle of Mediocrity,** which states that what happened in our solar system and on Earth is typical of the pre-biotic and biotic evolution anywhere else in the universe. In other words, Earth is not special.

For a variety of reasons, we can expect that other life is probably based on **organic molecules,** carbon-nitrogen-oxygen-hydrogen compounds. These elements are among the most abundant in the universe. Additionally, water (H_2O) provides an excellent solvent necessary for introducing compounds into and removing waste chemicals from living systems. Furthermore, carbon can build up long molecules, a requirement in the complexity of living systems. Carbon-based compounds can also be oxidized to yield the energy necessary to maintain life. Observational and laboratory study of a vast variety of environments suggests that all are locales where complex organic molecules can form through nonbiological processes. Such environments include pre-planetary nebula (as evidenced by organic molecules in carbonaceous chondrites), the primordial Earth atmosphere (replicated in laboratory experiments), comets, the atmosphere of Jupiter and the surfaces of its moons, interstellar clouds, and so on. In this circumstance, a major part of the path leading to the advent of life may be completed before a habitable environment on a planetary surface becomes available. On Earth, the primordial soup out of which life evolved may have been filled with all the necessary chemical requirements coming from the capture of comets.

Chemically, the only reasonable alternative to carbon is the element silicon. However, silicon is less common than carbon in the universe and has no effective solvent. Silicon can build long molecules, but it prefers the strong bonds of SiO_2 (sand). Also, silicon-based reactions are less energetic than carbon reactions. In other words, if carbon is present, silicon is most unlikely to be the basis for life. Exobiologists point out, however, that these considerations do not mean that other carbon-based life would look or act like terrestrial creatures.

Is there any evidence for life anywhere else in the universe? At present, the answer has to be a resounding no. And how would someone discover otherwise? This question is analogous to a situation in which two people live on an otherwise uninhabited island. One person's discovery of the other person's existence could happen in several ways. Scientists have classified these methods of discovery as type Ia, Ib, II, and III encounters, based on proximity. A **type Ia encounter** would involve remotely sensing some inadvertent sign of the other, such as the smoke from a cooking fire or detecting the broadcast signal for an *I Love Lucy* program from the 1950s. A **type Ib encounter** would involve remotely sensing a deliberate effort of the other to make his or her presence known—for example, the radio distress signal that was sent out by the Titanic shortly before it sank in 1912. Finding an artifact left by the other person (a beer can, or the Pioneer 10 and Voyager spacecraft, which have left the solar system) represents a **type II encounter.** Lastly, a **type III encounter** is an actual meeting (*Close Encounters of the Third Kind* in which friendly aliens arrange a meeting with us at Devil's Tower in Wyoming).

Philosophical Considerations

Multiple philosophical questions exist regarding the motivation to search for life elsewhere, particularly intelligent life, but these questions can only be noted here briefly. In support of the quest is, of course, the scientific and intellectual drive to understand everything in the universe as thoroughly as possible. It can also be argued that contact with another older, more advanced civilization that has survived would show humanity that it too can overcome the multiple problems of overpopulation, global pollution, political and economic conflict too often leading to war, and so forth.

Such a contact, however, may not be a positive event for humankind. In fact, the history of contacts between unequal civilizations on Earth raises the question of whether it would be at all desirable to have contact with a potentially more advanced civilization. In virtually every case on this planet, the meeting of a technologically superior culture with one less advanced has led to the (often catastrophic) demise of the latter.

SETI — The Search for Extraterrestrial Intelligence

Philosophical questions aside, what is the likelihood of finding another intelligent civilization? An answer lies in a number of factors, which have been expressed in the **Drake Equation**, an estimation of the fraction of stars with planets that harbor intelligent life. These factors include the fraction of stars that have planets, an estimation of which planets would have habitable conditions, and so forth: These are summarized in Table 16-1 with some indication of the pessimistic and optimistic estimations of the numerical values assigned to each factor. Currently, these factors are mostly guesses because sound scientific evidence to produce reliable numbers does not exist. (Note that as an alternative, the Drake Equation can be expressed slightly differently as a product involving the stellar birthrate function and the life span of civilizations):

Table 16-1: Factors Involved in the Drake Equation

Factor	*Plausible Lower Limit*	*Plausible Upper Limit*
Stars with planets	0.01	0.3
(only now are astronomers acquiring some evidence on the existence of planets around other stars)		
Stars with habitable conditions on one planet	0.1	0.7
(in **life zone** with temperature permitting liquid water, for example $M_{star} > 0.3M_{\odot}$; circular orbit, low obliquity, adequate gravity to retain atmosphere)		
Habitable conditions last long enough for life to evolve	0.1	1.0
Probability that life actually does evolve	0.1	1.0

(continued)

Table 16-1: ***(continued)***

Factor	*Plausible Lower Limit*	*Plausible Upper Limit*
Favorable conditions last long enough for intelligence to evolve	0.1	0.9
Probability that intelligence does evolve	0.1	1.0
Probability that intelligent life endures	0.0000001	0.1
Product = fraction of stars with planets that now bear intelligent life	10^{-14}	0.02
Probable distance to nearest civilization	3×10^6 pc	5 pc

Expressed in a different fashion, the pessimistic point of view suggests that intelligent life is extremely rare, with one civilization present in every 1,000 galaxies. On the other hand, the optimistic viewpoint is that there exists an intelligent civilization for every 50 stars in the Galaxy. Until significant advances are made in understanding planetary occurrence as well as evolution of biological organisms, these two limits cannot be narrowed.

One other matter should be mentioned: **Fermi's Question,** "Where are they?" The late Italian physicist Enrico Fermi simply pointed out that assuming a civilization has reached the technological ability to explore space, and further assuming that they will, then the Galaxy must have already been explored. Present human technology would allow expansion out into the Galaxy at a velocity of approximately 100 km/s, equivalent to a travel distance of 1 pc in 10,000 years. This is too slow for human beings to travel to nearby stars, but spacecraft could be sent and discoveries returned (slowly) to Earth by radio communication.

Present technology could support construction of a sophisticated exploring system, a **von Neumann machine,** which would not only explore another solar system, but at arrival would first search out construction materials (in an asteroid belt) and energy sources (solar energy; hydrocarbons from atmospheres of gas giant planets) to replicate itself, sending the next generation of exploring spacecraft to other stars. Under these circumstances, the time to cross the Galaxy is approximately 300,000,000 years, an extremely short time compared to the actual age of the Galaxy (and even shorter if advanced technological discovery would allow travel at much higher velocities). It could be argued that civilizations choose not to explore or perhaps do not survive long enough to move outward into the Galaxy. Even so, it would take only one civilization to do so to leave evidence of its visitation in the solar system. Yet there is no evidence that the solar system ever has been visited now or in the past (in particular, no report of an unidentified flying object has ever been confirmed as an extraterrestrial visitation).

Given the above uncertainties and the expense of space exploration, the detection of radio signals from other civilizations has been the chosen approach for **SETI,** the **search for extraterrestrial intelligence.** The first effort in SETI occurred in 1960: Project Ozma involved 400 hours of radio telescope observation of two nearby solar-type stars. Since then, there have been several dozen other projects, but none with a positive detection. Projects have included all sky searches in specific wavelength regions of the radio spectrum as well as searches paying specific attention to a selected sample of stars, for example, solar-type stars. SETI investigation is also piggybacked upon other radio astronomical research by analyzing these signals to detect any form of artificial pattern superimposed upon the background of natural radio radiation coming from natural radio sources.

While SETI is a very small part of all astronomical investigation, the motivation for continuing the search is strong. New technology has greatly improved the sensitivity of radio receivers, making detection of ever weaker signals possible. Astronomical revision of ideas concerning stellar and planetary formation and especially the ability to detect the existence of planets suggest that habitable planets are more likely than once thought. Similarly, there are occurring major revisions of ideas concerning the origin and development of life.

At the same time, humanity has both deliberately and accidentally signaled its existence. After its refurbishment in the early 1970s, the Arecibo radio telescope was used to send a radio signal toward the globular cluster M13. Any civilization living in or near its million stars might receive that signal in about 13,000 years, if it has a radio telescope tuned to the correct frequency pointing in the direction of Earth. The Pioneer 10 spacecraft, which left the solar system, carries a plaque with schematics showing the solar system, human beings, and the position of the Sun in the Galaxy. Voyager 2 carries a phonograph record with classical and rock music, human conversation, and other artificial and natural sounds. This record also was encoded with information to reproduce pictures of typical earth scenes.

Made in the USA
Las Vegas, NV
14 April 2025

20981109R00164